Contents

Section Three — Food Processes

Section Four — Marketing and the Environment

Published by Coordination Group Publications Ltd.

Editors:
Claire Boulter, Katie Braid, Rosie Gillham, Karen Wells, Sarah Williams.

Contributors:
Marion Brown, Angela Nugent, Jane Towle.

Proofreading:
Ellen Bowness.

ISBN: 978 1 84146 390 2

With thanks to Laura Stoney for copyright research.

With thanks to iStockphoto.com for permission to reproduce the photograph used on page 27.

© Crown copyright material on page 39 is produced with the permission of the Controller of HMSO and Queen's Printer for Scotland.

With thanks to Assured Food Standards for permission to reproduce the Red Tractor logo used on page 79.

Every effort has been made to locate copyright holders and obtain permission to reproduce sources. For those sources where it has been difficult to trace the copyright holder of the work, we would be grateful for information. If any copyright holder would like us to make an amendment to the acknowledgements, please notify us and we will gladly update the book at the next reprint. Thank you.

Groovy website: www.cgpbooks.co.uk
Printed by Elanders Hindson Ltd, Newcastle upon Tyne.
Jolly bits of clipart from CorelDRAW®

Based on the classic CGP style created by Richard Parsons.

Product and Market Analysis

Manufacturers don't often develop a brand new product — they usually redesign an existing one.
First, they do a product analysis — on their own or a competitor's product — to find ways to improve it.

Start with **Disassembly** and **Packaging Analysis**

Disassembly means taking a product apart and examining the bits. When you do this, take a photo of the packaging and food before you start. And remember to make notes. Write about:

1) The measurements of the product — make a table with the weight of each ingredient in. E.g. If you're disassembling a cheese and tomato sandwich, weigh the cheese, tomato and bread. This will give you the proportions of each thing.

Bread	243 g
Cheese	61 g
Tomato	120 g

There's about four times the weight of bread as cheese.

There's almost twice the weight of tomato as cheese.

2) The textures and colours of the various parts of the product, e.g. "It has a flaky golden crust". Describe the texture using words like dry, moist, crunchy, creamy, etc.
3) How the product is put together and how you think it was made, e.g. "The cheese is added last".
4) How it tastes, smells and looks. Be specific, e.g. "It's very bitter" (not "It's horrible").

The packaging is also useful — it shows you more detail about the product.

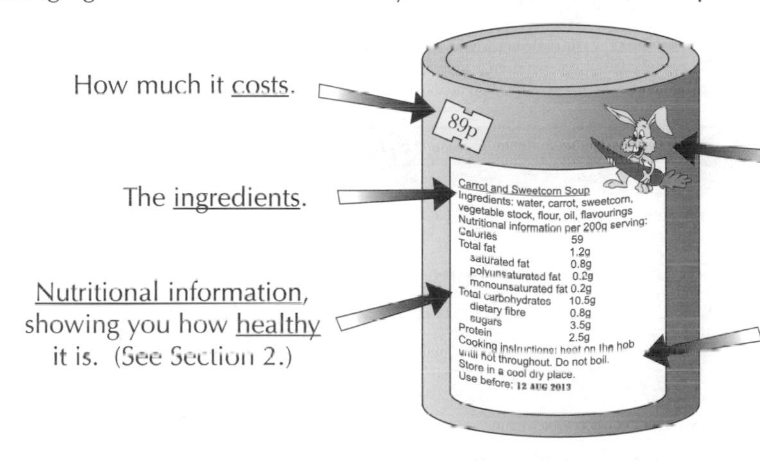

Clues from the style about the target market, e.g. using cartoons to target young kids.

Storage and cooking instructions, which tell you where to keep the product and how to prepare it.

Use the Info to Make **Your** Product **Better**

After you've analysed a product, decide its faults and find ways to make yours better. Think about:

1) The quality, quantity and proportions of the ingredients.
2) The size, shape, weight, appearance, texture and flavour.
3) The quality and effectiveness of the packaging.
4) The price — if you think it's too expensive for what it is, say why, and by how much.
5) The nutritional value.

When you write out how to make it better, be clear and exact.
E.g. If you reckon the original looks a bit pale and unappealing, don't just say "make it look nicer" — say something like "make sure the cheese topping is golden brown in colour".

Step one: work out how you can improve existing products

Ah, there's nothing like being critical. And there's loads to think about and try to improve — maybe the food could be cooked or stored differently, or perhaps it should be aimed at a different target market...

Product and Market Analysis

Before you design anything, you need to find out what people want. This information is dead important in helping you decide what your product should be like. Give the public what they want — simple as that.

Decide Who Your **Target Group** is

Even the very best products aren't everyone's cup of tea — some people like them and some don't.
Your target group is the group of people you want to sell your product to.
You should ask that group of people what they want the product to be like.

You can group people by things like age, gender, job, hobbies, lifestyle, income, or anything else — it'll probably be a combination of a few of these things.

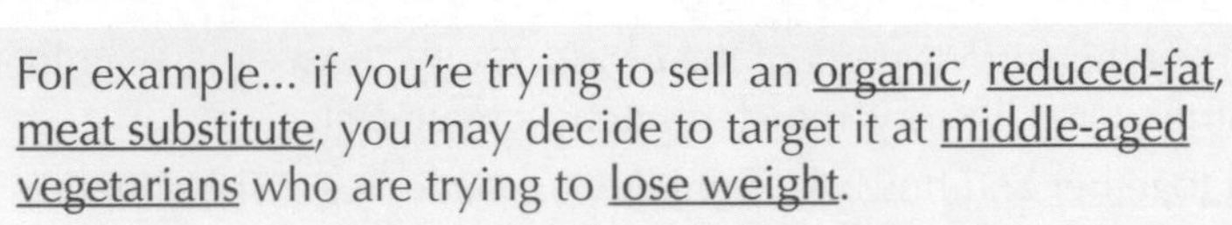

For example... if you're trying to sell an organic, reduced-fat, meat substitute, you may decide to target it at middle-aged vegetarians who are trying to lose weight.

But if it's a caviar-covered, organic, free range steak, you'd probably aim for rich, ethically-minded meat-eaters.

Think Carefully About What You Want to **Find Out**

Once you've decided on your target group, you need to decide what to ask them. You could find out:

1) Some information about the person answering your questions.
This could help you make sure they're within your target group, or give you extra info.
   - Are they male or female?
   - What age bracket are they in? (11-15, 16-20, 21-25 etc)
   - What job or hobbies do they have?

2) Do they already buy the kind of product you're thinking of developing?

3) Do they like a particular flavour or colour?

4) When and where do they buy it and where do they consume it?
This could affect the packaging you use.

5) Will they want to buy your version of the product?
Explain the advantage of your product over existing brands — would that be enough to tempt them to buy your version?

6) Is there something they would like from your product that existing brands don't have?

There are different types of questions you can ask to get this information — see p. 3.

Tailor questions to get the most useful information from your target market

So first off you need to decide who you want to sell your product to, then work out what info you need to know from them to help improve your product. Make sure your questions are relevant — there's no point asking what someone's favourite ice-cream flavour is if you're trying to design a new type of curry.

Market Research

So you've got your target group and decided what info you need from them.
Next, you need to phrase those all-important questions...

Questionnaires are Forms for People to Fill In

When you write a questionnaire, you should include:

1) A title — for example it could be 'Questionnaire Researching Favourite Puddings'.
2) A brief explanation of the purpose of the questionnaire.
3) A mixture of question types and not too many questions, so people don't get bored and give up answering them.

There are three basic types of question:

1) Closed Questions — these have a limited number of possible answers, e.g. do you like puddings? Analysing the results is easy for this type of question, e.g. by using graphs or charts.
2) Open Questions — these have no set answer, e.g. why is that your favourite pudding? They give people a chance to provide details and opinions. This type of questioning is more time-consuming and it's harder to draw conclusions from the results. But you could gain valuable information.
3) Multiple choice questions — these give a choice of answers. Sometimes the person answering can pick more than one.

You could include images and ask people which product looks most attractive.

Q4. What kind of puddings do you like?
Chocolate puddings ☑ Ice cream ☐ Fruit salad ☑

Interviews are Face-to-Face Conversations

1) For interviews, you can start off by asking the same sort of questions as in questionnaires — but then take the opportunity to ask follow-up questions, based on the answers you get.

2) Get your interviewees to give you extra information to explain their answers — this might help you get more ideas for your product. E.g. if their favourite pudding is trifle, ask them why they like it.

3) Interviews can give you more detailed information than questionnaires — you can have short conversations with people you're aiming to sell to. Just make sure you stick to the point.

4) A problem with interviews is that it's sometimes more difficult to analyse the results than with questionnaires (see p. 6 for more on analysis) because you might have asked different people different 'follow-up' questions.

Use a mixture of question types to get the information you need

Each of the different question types has pros and cons, so you'll get the most useful data by using a variety of question types. In interviews you can adapt or change your open questions depending on responses to previous questions.

Market Research

It's all very well asking people what foods they generally like or don't like, but when you're developing a new product there's no replacement for having people taste it and tell you what they think.

Use **Sensory Analysis** to Find Out **What People Like**

Sensory analysis is tasting samples of food and rating how good they are. Manufacturers ask consumers to do it, to find out what they think about new or existing products. This helps the manufacturer decide what characteristics their new product should have. There are different types of test:

1 Ranking or Rating Testing

People are asked to rank a number of similar products, or give them a rating:

Ranking Test	Name: David Smith
Taste the samples and place them in order of preference	
SAMPLE CODE	ORDER OF PREFERENCE
SPE12	2
SPE14	1

Rating system using symbols

☺ ☺ 😐 ☹ ☹ Circle the appropriate symbol

Hedonic Scale
1 = Hate
2 = Dislike
3 = It's OK
4 = Like
5 = Love

2 Star Diagrams

Testers rate the main characteristics of a product on a scale of 1-5. Each leg of the diagram represents a characteristic. The marks are then joined up, showing which aspects people like and which they don't.

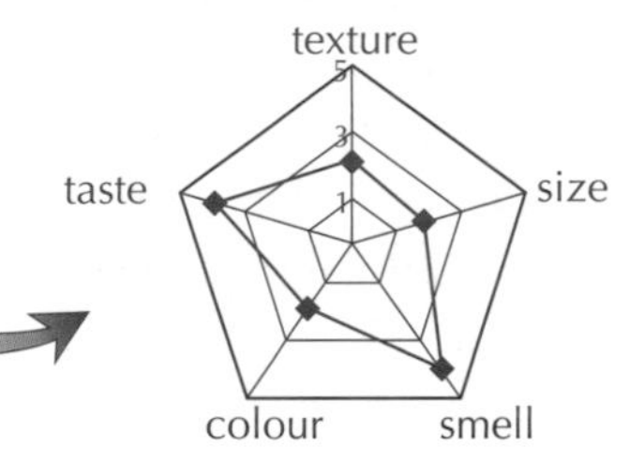

3 Triangle Testing

- This is when testers are given three samples and asked to say which is the "odd one out".
- Manufacturers use it if they're trying to develop a cheap or low-fat version of a food that tastes the same as the original. People taste two samples of the original and one of the new version — but they aren't told which is which.
- If most of the testers correctly pick out the new version, you'll need to re-design the product. But if they can't work out which one is different, you know you've designed a good alternative.

Do Your Sensory Analysis **Properly**

You need a group of people to be testers — ideally people from the target group.

1) Use a quiet area and give tasters water to sip to separate the tastes of different products.
2) Use small amounts of food and clean spoons. Don't let people put used spoons in the food.
3) Use codes or symbols for the products, to make sure the tasters aren't influenced by the name.
4) Make sure the tasters understand what they're meant to do.

Ranking, rating and triangle testing give data that's easy to analyse

You might get an exam question about how to carry out sensory analysis testing, interviews and questionnaires on a product, so make sure you know the ins and outs of doing each one.

Design Criteria

The process of designing and making something is called 'the design process' (gosh). The whole process can take a while, so it's usually broken down into smaller chunks.

*The **Design Process** is Similar in **Industry** and **School***

The things you'll have to do for your project and for the design question in the exam are pretty similar to what happens in industry. Remember:

- The best products are those that address a real need.
- That's why companies spend so much time and money on consumer research. The more people there are who would actually use a product, the more chance the product has of being a roaring success.

The rest of this section describes a typical design process.
It shows the sort of thing that happens in industry every day.

You need to understand the overall process, even though you probably won't have to actually do every bit of it.

*It Starts With a **Design Brief***

The design brief explains why there's a need for a new product. It usually includes:

1) an outline of the context (background) and who it involves (the target group)

2) what kind of product is needed

3) how the product will be used

The Design Brief is short and to-the-point — it's basically a starting point for the development of the product.

A typical design brief might say something like:

DESIGN BRIEF FOR CHILDREN'S BREAKFAST CEREAL

No currently commercially available breakfast cereal has a woodland animal theme.

We want you to design a product using this theme. A successful product will be tasty and nutritious, and will appeal to both children and their parents.

The design brief will help you decide which products to analyse and what you need to find out from your market research (see pages 2-4). You need to work out what will appeal to your target group — in this case you might talk to children and their parents about what they want from a breakfast cereal.

The design brief is the starting point for developing a new product

Make sure you're familiar with what a design brief includes and how a product is developed from one. You'll be given a design brief in your exam — so read it carefully and make sure your design covers everything that's mentioned.

Design Criteria

You've got your design brief and (hopefully) lots of initial ideas. That must mean it's time to work out what all your research shows and come up with some design criteria for your product...

Do Research to Draw Conclusions

When you've done your research you should use it to help with your design.

1) Summarise what you've found out — pick out the most important and useful findings, e.g. "hedgehogs are the most popular woodland animal with children".
2) Explain what impact the research will have on your designs, e.g. "the cereal will have a hedgehog theme".
3) Suggest ways forward from the research you've done, e.g. "one idea would be to make each piece of cereal hedgehog-shaped".

You Need a List of Design Criteria

1) The conclusions from your market research should show what kind of characteristics your product needs to have.
2) These requirements are your design criteria. (A list of design criteria is sometimes called a design specification.)
3) Each point says one thing about what the product should be like, e.g.

Design Criteria for a New Cereal
- made from oats
- contains dried strawberry pieces
- animal shaped cereal

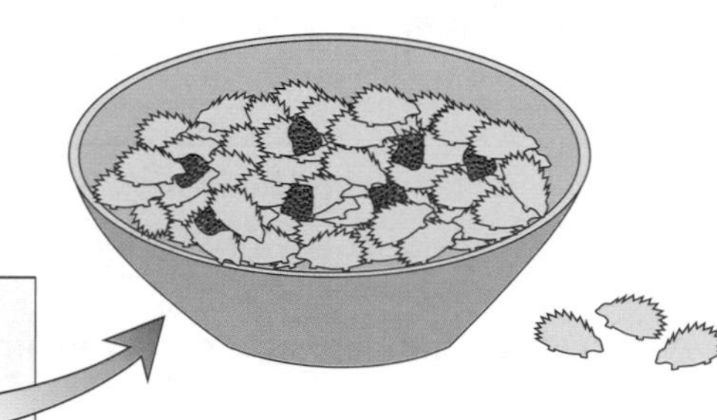

EXAM TIP
You'll be given a Design Brief and/or a Design Criteria in the exam — make sure you read them really carefully and highlight the key points.

4) You don't have to be really exact at this point — that comes later (see page 9). Just a few words for each point is enough.
5) But you do need to show how the criteria are related to your research — e.g. it's fair enough to have "the omelette should be hexagonal" as one of your criteria, but only if your research analysis concludes that people want a hexagonal omelette — don't just make it up because it sounds interesting.

Use your research findings to come up with some design criteria

Making a list of your design criteria will help rule out ideas that aren't suitable. The more criteria you have, the more precise you can be about how your product will turn out, but don't go mad and lose the flexibility to change your design.

Generating Proposals

Now hold on to your hats — this is the creative bit. Time to start generating lots of ideas.

There are a few **Tricks** that can help you **Generate Ideas**

1) Work from an existing product or recipe — but change some of its features or production methods so that it fits in with your design specification, e.g. the colour, size, ingredients, etc.
2) Think about the functionality of the product — how well it meets the needs of your target group. You also need to think about how well your idea meets the design brief and specification.
3) Think about how to get the aesthetics you want — how your product looks, smells and tastes like. How will you get the texture you want? What about the nutritional content?
4) It may help to do a spot of brainstorming...

Brainstorm to Produce **Initial** Ideas

1) First, think up key words, questions and initial thoughts about your product. Write down the design criteria and research conclusions too.
2) Don't be too critical at this stage — let your imagination run wild. Even if an idea sounds ridiculous, put it down anyway.
3) Be creative and get as many ideas as you can. Afterwards, decide which ones are good (and so are worth developing) and which ones just don't work.
4) Use word association — choose a product and write down any related words. E.g. biscuits, soft, straws, melt and blue are all associated with... cheese.

EXAM TIP
You could get a question asking for initial ideas — just do them as freehand sketches, and if it's worth 5 marks, include 5 details.

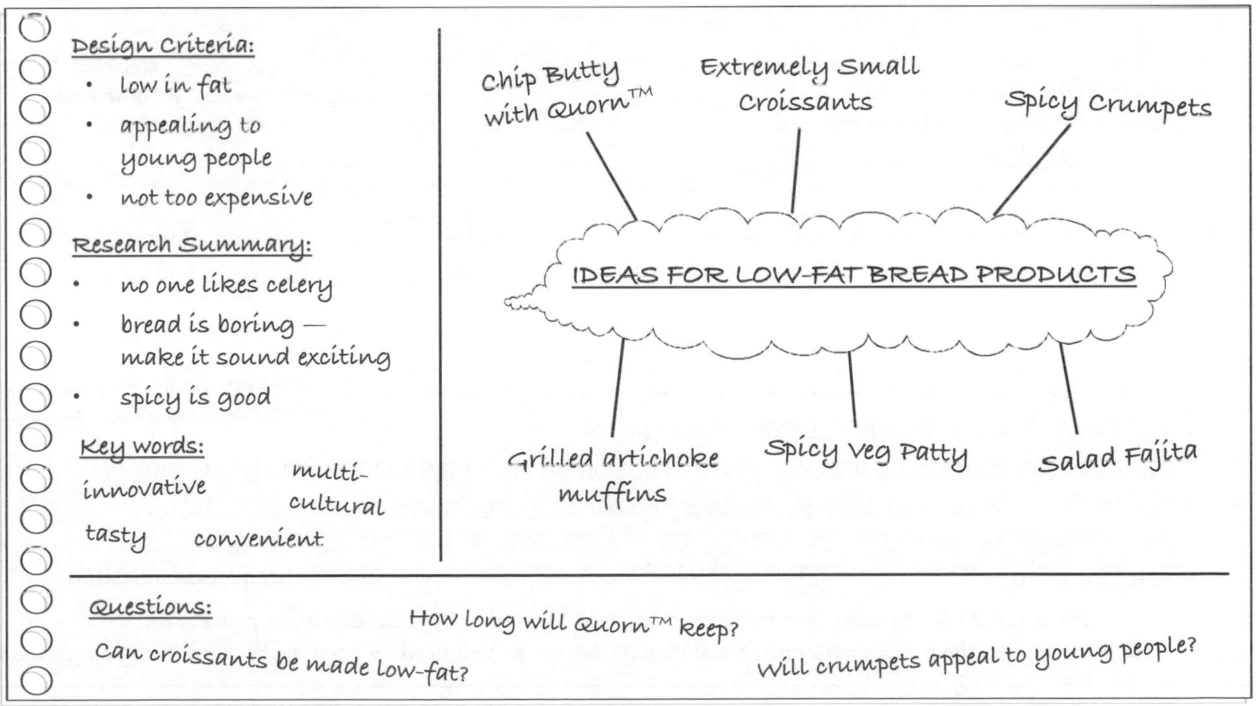

Design Criteria:
- low in fat
- appealing to young people
- not too expensive

Research Summary:
- no one likes celery
- bread is boring — make it sound exciting
- spicy is good

Key words:
innovative
multi-cultural
tasty
convenient

IDEAS FOR LOW-FAT BREAD PRODUCTS
Chip Butty with Quorn™
Extremely Small Croissants
Spicy Crumpets
Grilled artichoke muffins
Spicy Veg Patty
Salad Fajita

Questions:
How long will Quorn™ keep?
Can croissants be made low-fat?
Will crumpets appeal to young people?

Generate ideas by looking at existing products or brainstorming

Try to think of a range of ideas that are really different from each other. Then you might be able to combine the best features from a few of them and get the most delicious meal since fish and chips.

Generating Proposals

So you've come up with a load of ideas and used your design criteria to narrow it down to a few — now it's time to learn how to present your ideas to clients, potential investors and, of course, examiners.

You need to Come up with a **Range of Designs**

1) Once you've sorted out the good ideas from the bad ones, annotate (i.e. add notes to) each good design idea to fully explain what it is and why it's good. You could mention:

- materials
- sizes
- user
- shape
- cost
- advantages/disadvantages

Your notes should link these features to your design criteria.

2) You need to produce a range of different solutions — about 3 or 4 — that meet the design criteria.

3) It's also important that you think you could actually make them — don't go overboard on exciting ideas that you could never produce for real.

Present Your Ideas **Clearly**

1) To present your ideas, it's usually best to keep it simple — a freehand sketch will do fine, as long as it's clear.

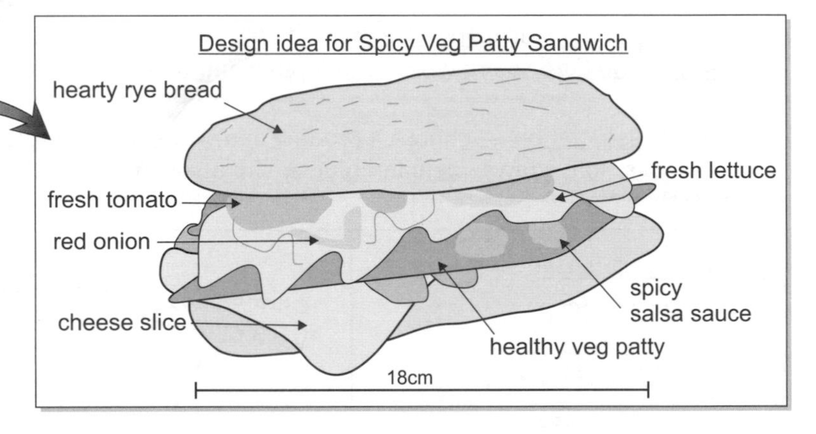

2) Once you've got a few possible designs, have a good think about them all, and decide which is the best idea — i.e. the one that most closely matches the design brief and design criteria. This is the one you should develop.

3) You should check that the nutritional content of your chosen product meets your design criteria too, e.g. if it should be low in fat, check that it is. You could do this using nutritional analysis software:

See page 62 for more on using computers to help with designing.

- The software creates a computer model of your product, from your recipe.
- It calculates the nutritional content of your ingredients and any nutritional losses due to cooking.
- It also tells you the recommended guidelines for your target market, so it's easy to see if you need to adjust your recipe to fit your design brief. Remember, any changes you make will affect the product in other ways too, e.g. the taste.

It's a good idea to carry out nutritional analysis throughout the development process too. If you modify the recipe, the nutritional content will change too — so you need to keep a check on it as you go along.

Practise drawing design ideas for different criteria...

... it'll help you to get the hang of quickly coming up with design ideas. This page is really important — you'll have to do this stuff as part of your project work or exam. So make sure you've taken it all in before you move on.

Product Specification

Once you've picked out the best idea to develop, you're ready to put together a product specification.

The **Product Specification** Describes the Product

1) The product specification expands on your chosen idea and says exactly what the product is, not just what it tries to do.

2) It describes what the product contains, how it looks and tastes and so on. It should have exact figures and measurements.

3) In your product specification, include some or all of the following:

- how it will look
- how it will taste
- how it should be stored
- size and weight
- safety points
- costs

EXAM TIP
Your product specification does NOT say how to make it — that goes in the Manufacturer's Specification (see pages 13-14).

4) Put your specification together using bullet points, rather than wordy explanations.

- Each sandwich will weigh 180-200 grams.
- Manufacturing cost will be under 75p per unit.
- It will use brown bread.
- It will have a ham, cheese and cucumber filling.
- The primary flavour will be cheese, with a spicy after-taste.

Use words like "will", "should" and "must" in your specification.

Make Sure it's **Realistic**

1) For your project, chances are you'll have to actually make the product according to the product specification — so all your requirements need to be things you're capable of producing.

2) Once you've got a product specification, you'll need to compare it to the design criteria and confirm that each point is satisfied.

Some points will be harder to check than others. For example, if one of your design criteria is "must be very sweet", you'll have to actually make the product before you can check (by tasting) whether it's sweet enough. It's a good idea to take photos of any taste testing you do.

Make sure you know what goes in a product specification

If I told you that product specifications were going to get your pulse racing, I'd be lying. To be honest, they're a bit dull. But they're a vital step in designing and manufacturing a new product. So make sure you know how to do them.

Product Specification

Let's say you've come up with the ultimate design idea for your food product — something that's not been thought of before but you reckon will be dead popular. Peanut butter and chocolate bagels, that sort of thing. There's nothing to stop someone else from taking your idea, developing the product and making loads of money themselves — or is there...

Protect Your Design Ideas so YOU Benefit From Them

1) If you design a food product that's original (never been done before) then you can legally own your idea, just like you own physical property — it's called intellectual property (IP).

2) You can register different features of your design idea as intellectual property. This means these features are protected — they can't be stolen by anyone else.

3) If your idea is protected, it gives you exclusive rights to develop your product and hopefully go on to make tons of money from it. If someone else wants to do the same thing, they need your permission and they have to pay you for it.

4) If you don't protect a design idea that turns out to be successful, anyone else can copy your idea and benefit from it.

EXAM TIP
Try to write one point for each mark in the exam question — e.g. six points for six marks.

5) You can register the shape, colour, texture and all kinds of other stuff about your design idea. This can include a new ingredient you're using, the recipe, the production process or the product packaging. E.g. In 1937, Coca-Cola registered the design for their cola bottle — so no-one else could copy it.

So if you invent an exciting new flavour of ice-cream, made using a new technique...

...come up with a name and an innovative way of packaging it...

...and design a logo for your company...

...then you can register the flavour, production method, recipe, packaging design, name, logo and anything else that you've come up with.

6) You can only protect your ideas for about 20-25 years — after this, anyone can develop them.

You can protect all kinds of different features of your product

This stuff is well worth knowing about — think how frustrating it would be to come up with a killer idea that's guaranteed to make you a millionaire, then have someone else steal it and get all the glory...

Development

Once you've put together your product specification, it's time to give your product a try. Yep, actually make it — and then make improvements to it.

You can **Develop Your Design** in Different Ways

Depending on the type of product that's being produced, there are a few ways you can develop it.

1) You could make some more detailed sketches. This might help you decide on some of the smaller details you hadn't thought about before, e.g. how the different toppings of a pizza would be arranged.
2) Do some practical experimentation with different aspects of the design. You could:

   - change an ingredient, e.g. soft brown sugar instead of caster sugar
   - change a component, e.g. pears instead of apples
   - change the equipment, e.g. mix by machine instead of using a spoon
   - change the process, e.g. add the cheese last, before baking.

   Trying out different versions of your design is called modelling — and each different version is a model. (See below for more on modelling.)
3) Use your target group's opinions about developments to help you give them what they want.

Make **Changes** and **Compare** Models

If your product is for freezing and reheating, you'll need to try doing that — and then evaluate the product after reheating.

1) After you've made the first real model of your design idea (called a prototype), you need to do some tests to check it's how you wanted it to be — this is called evaluation.
2) These tests could cover appearance, texture, taste, smell and other things. Check it against all the criteria in the design specification too.

It's dead important that your sensory analysis (see page 4) and other tests are thorough and rigorous — you need to be super-critical of your model so that you can make the final product as good as possible.

3) You'll probably find there are some things in your initial model that didn't work out the way you'd hoped — maybe it tasted great but was really expensive, in which case you could try using some cheaper ingredients or making a smaller product.

The sponge cake was too greasy, so in the next model I'm going to try using butter instead of lard.

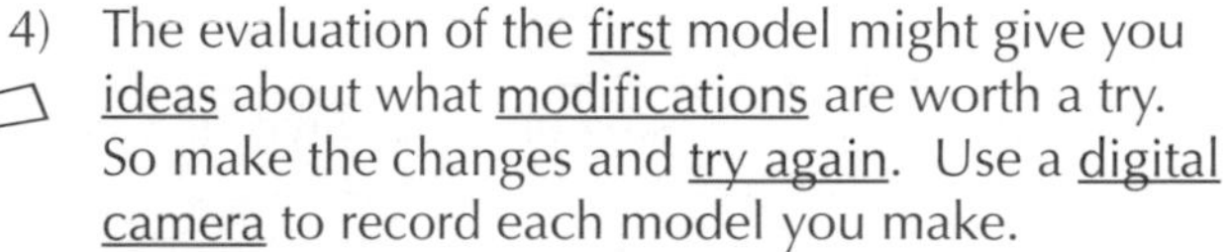

4) The evaluation of the first model might give you ideas about what modifications are worth a try. So make the changes and try again. Use a digital camera to record each model you make.
5) Put each model you make through the same tests. That way you can compare them fairly and see if you've actually improved things.

Development is a vital part of the design process. Ideally you should solve all the potential problems with your design at this stage.

Conduct tests and learn from them

Modelling and evaluation go hand in hand. It's pointless baking a cake and eating it if you're not going to bother learning anything from it. Well, OK, it might be fun, but you're not here to have fun. You're here to learn.

Development

You've got to keep testing and adjusting your product until you're left with the best version of your best design.

Keep Going Until You Get it Just Right

You might have to modify quite a few aspects of your design. Developing a product can involve trying out lots of different changes (see page 11).

Changing one thing might mean you need to change something else.

For example, say you bake a cake in a wide, circular tin instead of a deep, loaf-shaped tin. The cake will now be thinner and could burn more easily... so you might have to alter the cooking time or temperature.

Here's a summary of how development works — every time you try something new:

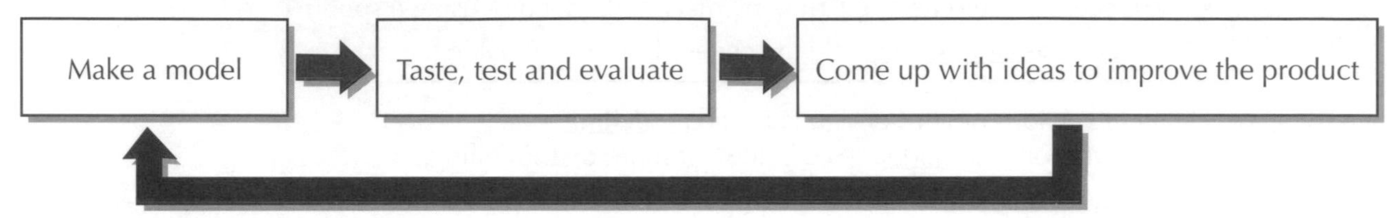

1) You might have to keep changing your product to make sure you meet the design specification. That's fine — the whole point of development is to find out what works and what doesn't.
2) But you can't change the design specification — because then you probably wouldn't be meeting the design brief any more.
3) In other words, you've got to make your product fit the design specification, not the other way round.

Remember to note down what you're changing and why. This shows you're doing things properly.

Part of your design brief might be to make your product environmentally-friendly — so you'd need to think about these kind of things:

1) Are your ingredients from a local source? (This helps to reduce pollution because they don't travel as far.)

2) Are your materials sustainable? (This means you're not using up natural resources until they run out.)

3) Can you reduce the amount of waste food and materials? (Think what you can reuse before throwing it away.)

4) How much energy is used to make your product? (Reduce energy used, or use renewable energy.)

See pages 78-79 for more on environmental issues.

Model, taste, test, evaluate, improve — repeat until satisfied

For every change that you make, remember to keep careful notes of exactly what you changed, why you changed it and how the final product turned out. That way when you come up with a perfect design, you'll know how you got there.

Manufacturer's Specification

When you know exactly what you're going to make and how, you need to communicate all that info to the person who's actually going to make it — the manufacturer. (OK, so chances are that you're the manufacturer, but don't relax — you still need to do all this stuff, and it could be in the exam.)

You Need to Make a **Manufacturer's Specification**

A manufacturer's specification can be a series of written statements, or working drawings and sequence diagrams. It should include enough detail for someone else to make the product — stuff like:

1) how to make it — a clear description of each stage, which may include photos
2) a list of ingredients with precise amounts of each
3) the dimensions of the product, given in millimetres
4) tolerances — the maximum and minimum sizes or weights for each part, e.g. 'the water icing must be between 4 mm and 6 mm thick'
5) finishing details — detailed descriptions of techniques used for any toppings, decoration, etc
6) quality control instructions — when and how checks should be made (see page 63)
7) costings — how much each part costs, and details of any other costs involved

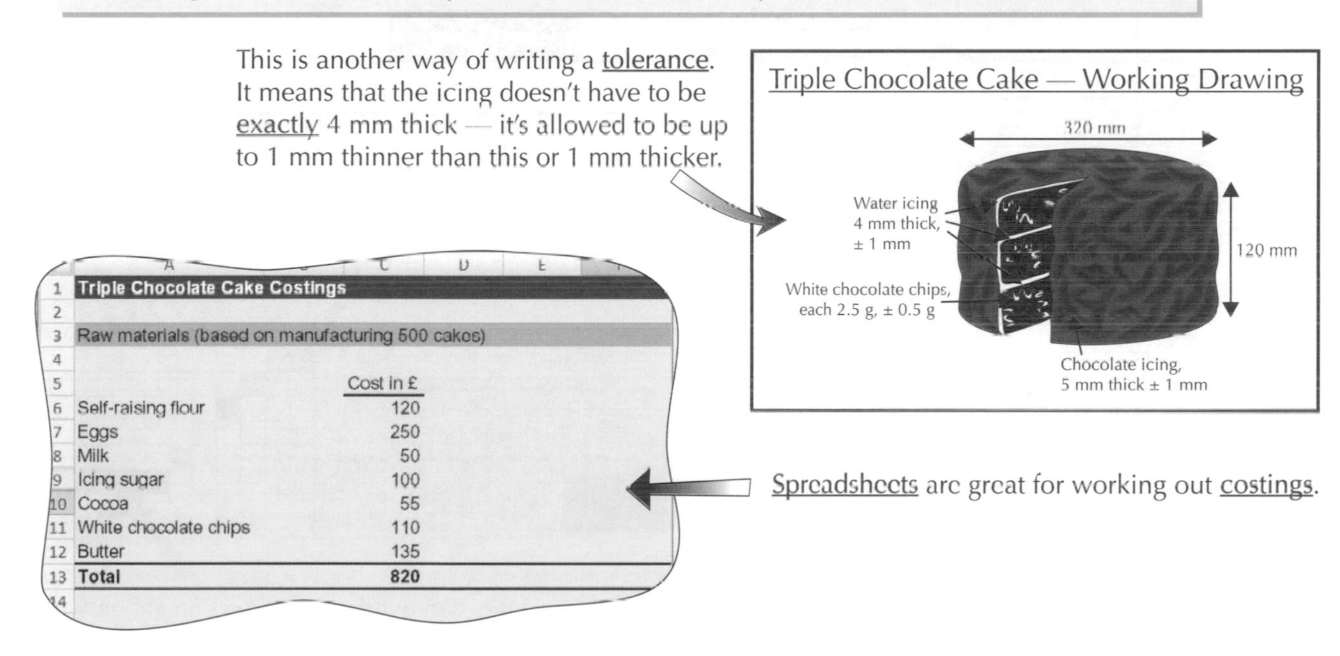

Triple Chocolate Cake Costings	
Raw materials (based on manufacturing 500 cakes)	
	Cost in £
Self-raising flour	120
Eggs	250
Milk	50
Icing sugar	100
Cocoa	55
White chocolate chips	110
Butter	135
Total	**820**

Plan How Long the **Production Process** Should Take

When you get to this stage of product development, you also need to plan:

1) any changes needed to make it suitable for mass-production
2) how long each stage will take
3) what needs to be prepared before you can start each stage
4) how you'll ensure consistency and quality

See the next page as well for some different ways to help with this planning.

A manufacturer's specification tells someone else how to make the product

The main thing to remember with a manufacturer's specification is that you have to be really specific and make sure you've covered every aspect of your design. Try giving it to someone else to make and see how they get on.

Manufacturer's Specification

Manufacturing a product takes a shedload of careful planning. Luckily there are some useful techniques you can use to help make the planning as straightforward and painless as possible.

Use **Charts** to Help You

You need to work out what order to do things in. It's also important to work out how long each stage will take and how these times will fit into the total time you've allowed for production.

1 Work Order

This can be produced as a table or flow chart. The purpose is to plan each task in sequence. You should include quality control checks.

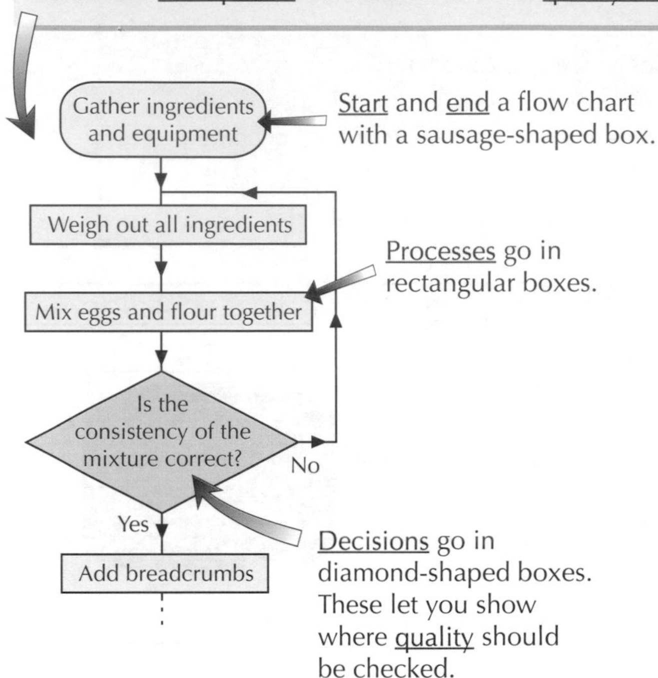

Start and end a flow chart with a sausage-shaped box.

Processes go in rectangular boxes.

Decisions go in diamond-shaped boxes. These let you show where quality should be checked.

2 Gantt Chart

This is a time plan. The tasks are listed in order down the left-hand side, and the timing plotted across the top. The coloured squares show how long each task takes.

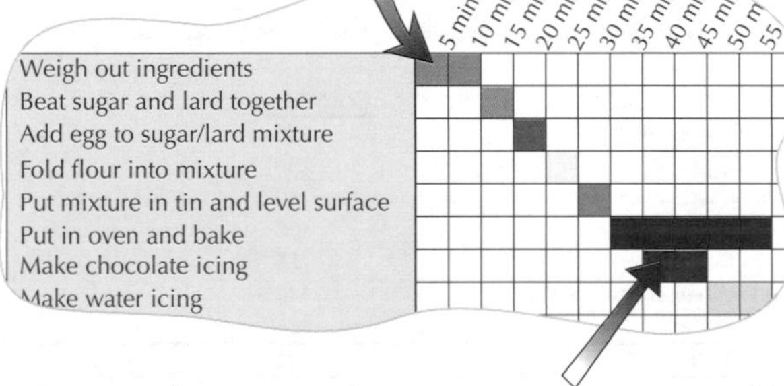

Some tasks can overlap, e.g. you can get on with making the icing while the cake is in the oven.

Test *The* ***Finished Product***

When you think you've got the final product, it's vital to photograph and test it. You have to make sure it meets the original design criteria (see pages 5-6).

More questionnaires or surveys may help here. Ask people to give their opinions about the finished product — people in your target group if possible.

Practise constructing work orders and Gantt charts

You might have to plan the manufacture of your product, so make sure you know how. Be realistic about how long things will take — there's no point claiming you can do something in 10 minutes if it's really going to take two hours.

Warm-Up and Worked Exam Questions

Make sure you've understood this lovely section by having a go at these warm-up questions.

Warm-up Questions

1) a) What is a target group?
   b) What features can you use to describe a target group?
   c) List five sensible things you could ask a target group if you were making a new kind of salad.
2) Imagine you are researching opinions about a new sandwich product. Write one example of:
   a) a closed question b) an open question c) a multiple choice question
3) Give three examples of features of a product that you could register as intellectual property.
4) Imagine you are making a new kind of low-cost pizza, but your first model's texture is too dry. Describe two ways you could try developing your product to improve the texture.
5) Draw a Gantt chart for making Spaghetti Bolognese.

Worked Exam Questions

You're unlikely to get through your GCSE without having to design a product. It can seem a bit daunting at first, so have a look at this worked example to get an idea of how it's done, then have a go at one yourself.

1 Design a new high-energy snack that meets the following criteria:

- contains a lot of carbohydrates and sugar,
- uses ethically produced ingredients, ← *See pages 78-79 for more about ethically produced ingredients.*
- can be eaten 'on the go'.

(a) Use sketches and notes to show your initial ideas for your chosen product. Do not draw the packaging. ← *Don't draw any packaging — you won't get any marks for it.*

(6 marks)

Design 1

Oat bar with chocolate topping

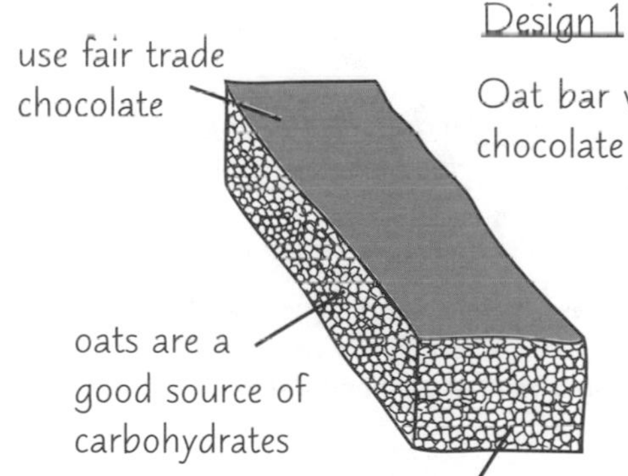

oats will be held together using golden syrup and margarine, so it won't crumble or fall apart — this will make it easier to eat 'on the go'.

You need to produce a range of ideas — but you don't have time to go overboard. Three is a good number.

Design 2

Cookie with dried fruit and nut pieces

small enough to fit in a pocket

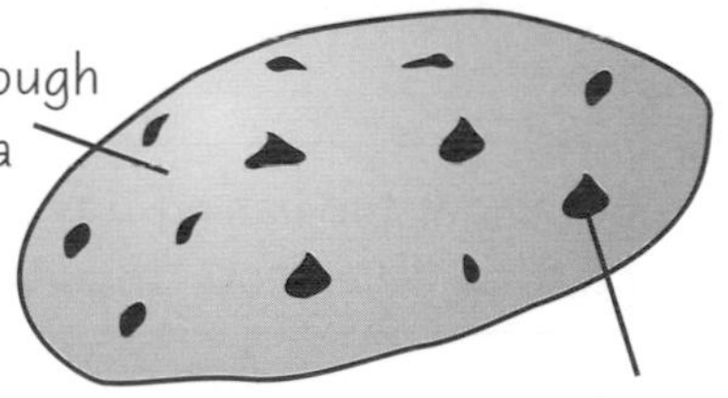

use fair trade dried fruit

To get the full 6 marks you need to annotate your sketches to explain your ideas and to say how they meet the design criteria.

Design 3

Summer fruit muffin

use locally-grown fruit

cake is a good source of carbohydrates and sugar

Make sure each design idea is different from the others — you won't get marks if they're too similar.

Check you've covered all the design criteria. It helps if you tick off each one as you go.

Worked Exam Questions

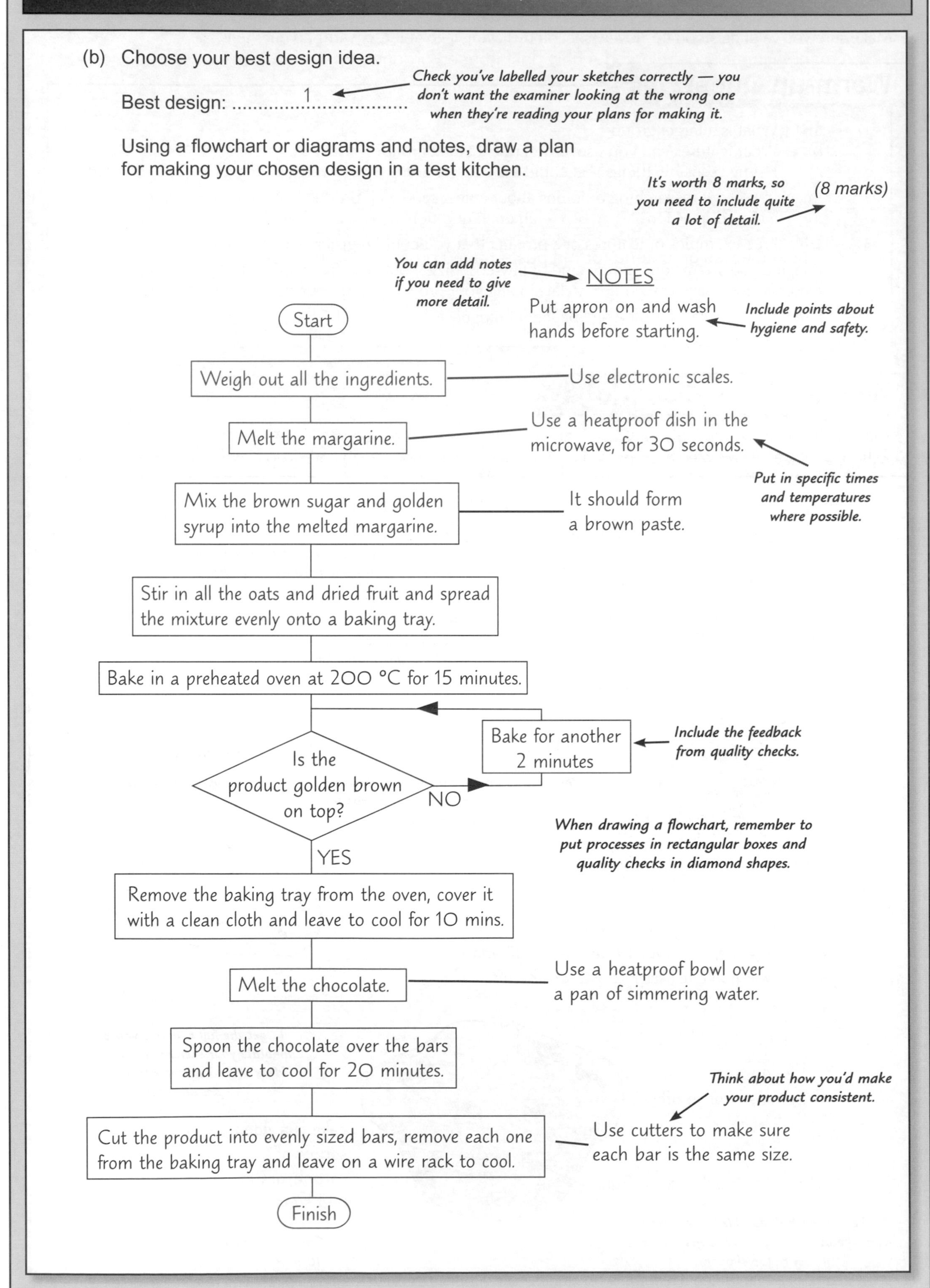

Exam Questions

1 A manufacturer is developing a new pasta-based product.

The product should:

- be a savoury meal for one person
- have a tomato-based sauce
- be high in carbohydrate
- be high in protein
- offer sensory appeal

(a) Use notes and/or annotated sketches to produce two different design ideas that meet the design criteria for the pasta-based product. Do not draw any packaging.

(10 marks)

Exam Questions

(b) Choose your best design idea.

Best design: ....

Using a flowchart or diagrams and notes, draw a plan for making your chosen design in a test kitchen. Include safety checks in your plan.

(5 marks)

Revision Summary for Section One

Congratulations, you've made it to the end of the first section. There's a fair bit to take in in this section, and it's not all that straightforward, so make sure you have a good grasp of it by having a crack at these revision questions. The answers can be found in the section, so if you get stuck on anything then go back and have another look, then try the question again. Keep going until you can answer them all without having to refer back to the page.

1) What does "disassembly" mean?
2) List three useful things you can find on the packaging of a food product.
3) Write down three pieces of information you might want to find out from your target group before you start designing a new product.
4) What is an open question?
5) Give one advantage and one disadvantage of interviews compared to questionnaires.
6) What's the point of sensory analysis?
7) Briefly describe three different types of sensory analysis.
8) Why do companies think consumer research is so important?
9) What information does a design brief give? Why is it important?
10) What are design criteria?
11) Describe two ways of coming up with ideas for a new product.
12) What should you do after you've come up with plenty of ideas?
13) What could you use to check the nutritional content of your product?
14) List three points that you should include in a product specification.
15) Why is it a good idea to protect your design ideas?
16) Can other manufacturers use your design idea if you have protected it?
17) What does modelling mean in Food Technology?
18) Why can't your design specification change during the development of a product?
19) The manufacturer's specification for a batch of fairy cakes says that each cake must weigh 52 g ± 3 g. Explain what this means.
20) How do each of these help your planning?
    a) a flow chart
    b) a Gantt Chart
21) What should you check when you reckon you've got the final product?

Carbohydrates — Sugar

Carbohydrates are one of the major food groups and there's plenty of stuff to learn about them...

Carbohydrates are Needed for **Energy**

Carbohydrates are split into three types: sugar, starch and fibre.

SUGAR

Includes simple sugars like glucose and fructose, as well as double sugars such as lactose and sucrose. They're easier to digest than starch.

STARCH

Starch is a complex sugar. It needs to be broken down by digestion before the energy can be used. That's why it's good to eat starchy foods like pasta and rice a few hours before playing loads of sport.

FIBRE

Fibre is another type of carbohydrate.
Bran, fruit, beans and brown bread contain lots of fibre (see page 30).

Here's some sciency mumbo-jumbo — it's all to do with the chemical structure:

simple sugars	=	*monosaccharides (the most basic sugar molecules)*
double sugars	=	*disaccharides (made up of 2 monosaccharides)*
complex sugars	=	*polysaccharides* } *(long chains of monosaccharides)*
fibres	=	*non-starch polysaccharides (NSPs)* } *(long chains of monosaccharides)*

Carbohydrates are good sources of energy. But energy that's not used is stored by the body as fat. So it's often carbohydrates, not fats, that make people overweight.

Several **Types of Sugar** are Used in Home Baking

1) Granulated sugar is for general kitchen use, e.g. to sweeten tea or breakfast cereal.
2) Caster sugar has finer crystals than granulated sugar. It's used for baking, especially cakes and biscuits, which need to have a fine texture.
3) Brown sugars — demerara and muscovado are brown sugars with strong, distinctive flavours. These are used in rich fruit cakes, gingerbread and Christmas puddings.
4) Icing sugar is a white, powdery sugar used for icing and sweets.

Most of these originally come from sugar cane.
Sugar also naturally occurs in things like fruit and honey.

Sugar, starch and fibre are all types of carbohydrate

Sugar is obviously good in some ways — it tastes great, and you get sweets, cakes, biscuits, chocolates and all things good from it. But if you have too much of it, it can rot your teeth and make you put on weight.

Carbohydrates — Sugar

Sugar isn't just used to sweeten cakes, it's used in all sorts of foods for all sorts of different reasons.

Sugar is Used in **Loads** of **Food Products**

Sugar is used widely in food manufacturing, even in savoury products.
Just look on some ingredients labels — fructose, dextrose, sucrose, inverted sugar, maltose, lactose and glucose are all sugars.
Sugar has lots of functions:

1) It makes things sweet (obviously) or 'softens' very sharp flavours, e.g. in lemony desserts.

2) It acts as a preservative, e.g. in jam.

3) In creamed mixtures, sugar is beaten with fat, which aerates the mixture (adds air to it) and helps lighten it, e.g. in cakes.

4) It speeds up fermentation, e.g. in bread.

5) Sugar adds colour, e.g. in cakes, biscuits and pastries.

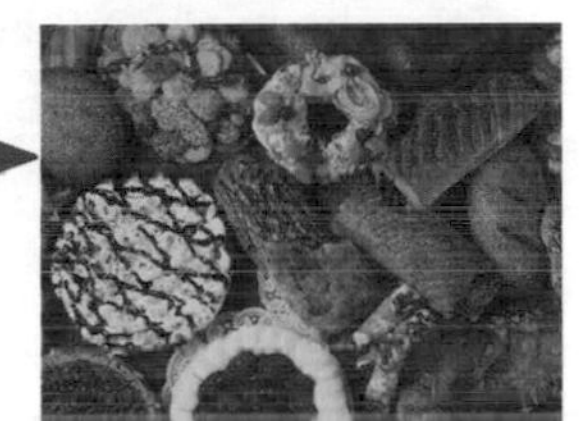

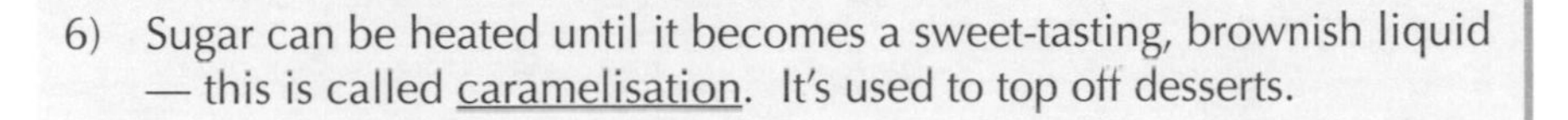

6) Sugar can be heated until it becomes a sweet-tasting, brownish liquid — this is called caramelisation. It's used to top off desserts.

Sugar **Substitutes** are Sometimes **Healthier**

1) Sugar substitutes can be used to sweeten drinks and foods.

2) They're better for your teeth than sugar and contain far fewer calories, so they're good for people who are on a slimming diet. They're also good for diabetics, who have to control their sugar intake.

3) Sugar substitutes shouldn't be used for home baking because they don't have the same properties as cane sugar, described above.

Learn the different reasons why sugar is added to food

It's amazing how much sugar is in food, and not just in the sweet stuff. Next time you eat some processed savoury food, have a read of the label and you might be surprised just how much is there.

Carbohydrates — Starch

Because of starch's properties, it has a variety of uses. (And not just in food — it's the traditional way to stiffen shirt collars, for instance. But I don't suppose the Food Tech examiner will be terribly interested in that.)

Starch can alter the Structure of Foods

STARCH IS USED AS A BULKING AGENT

Starch granules swell when a liquid is added, and so can provide the bulk of a product, e.g. the starch in flour makes up most of the volume of pasta.

STARCH IS USED AS A GELLING AGENT

When moisture is added to starch granules and heat is applied:

1) Starch granules begin to absorb the liquid and swell.
2) At 80 °C the starch particles break open, making the mixture thick and viscous. This is gelatinisation.
3) Gelatinisation is completed when the liquid reaches 100 °C.
4) The thickened liquid now forms a gel.
5) On cooling, the gel solidifies and takes the form of the container it's in.

E.g. custard is a thick liquid when first cooked, but becomes solid when it cools.

Starch is used to Thicken Foods

STARCH IS USED AS A THICKENING AGENT

Sauces and gravies are often made using starch (in flour) and liquid. The thickness depends on the proportions of starch and liquid.

1) The starch and liquid are mixed together.
2) The starch particles form a suspension — they don't dissolve.
3) The mixture is stirred to keep the particles suspended.
4) Heat is applied and gelatinisation occurs, which causes thickening.

STARCH IS USED IN MANUFACTURED PRODUCTS

Modified starch (see next page) is used to thicken things like instant desserts, whipped cream, yoghurts and packet soup. Usually a liquid is added to the starch and it is stirred or whisked.

Starch is used to bulk up, gel and thicken foods

Starch is pretty important stuff and it's used in loads of different types of food. Make sure you learn the difference between bulking, gelling and thickening, and that you can give an example of a type of food for each one.

Carbohydrates — Starch

Starch has many different uses in food manufacture, and it can be modified to make it even more useful. Starch isn't the only nutrient that can affect a food's properties though — for example, bread relies on air and gluten to give it its texture.

Modified Starches** are Called **Smart Starches

You can get some starches that have been treated so that they react in a particular way in certain conditions. They're known as modified starches or smart starches.

1) Pre-gelatinised starch thickens instantly when mixed with hot water, e.g. packet custard, instant noodles.

2) When protein is heated it can coagulate (become more solid) and squeeze out the fat and water. This is called syneresis. Some starches allow products to be reheated with no syneresis. This is handy with frozen foods (e.g. lasagne) so that they keep their moisture and nutrients when they're cooked.

3) Normal starches can be affected by acid, so that they don't work properly. But some modified starches are immune to it, so they can be used to thicken acidic products, e.g. salad cream, which contains vinegar.

Gluten** helps Bread Dough to **Stretch** and **Rise

Bread contains lots of starch, but there are other important nutrients that give bread its properties. To make sure a loaf of bread doesn't turn out like a heavy doorstop, you need the dough to be elastic.

1) When dough made with flour is kneaded, a protein called gluten is formed.

2) To get a well-risen loaf of bread it's best to use strong bread flour because it will form more gluten than other types of flour.

3) Gluten gives dough its elasticity (stretchiness) and helps bread to rise.

4) The dough mixture contains yeast, which ferments the sugar to produce carbon dioxide — a gas.

5) The gluten stretches to hold the carbon dioxide — this is what makes bread rise.

6) When gluten reaches a high temperature it coagulates (it changes into a more solid state). The dough stays stretched to give the light, airy texture of well-risen bread.

Starch can be modified to change its properties

Modified starches are really useful. For example, modified starch from barley has similar properties to fat, and can be used to make stuff like low fat cakes or biscuits. Handily, some types of modified barley starch also freeze really well.

Proteins — Meat, Poultry and Fish

Meat, poultry and fish provide high-grade protein and other essential nutrients. But bacteria also like them, so you have to be really careful when buying, storing, preparing or cooking them.

*Protein is Needed for **Growth** and **Repair***

1) Protein helps our bodies to build and repair muscles, tissues and organs, and helps children grow.
2) Protein is made of amino acids. Your body can make some amino acids but not others. You have to eat the amino acids that your body can't make — the essential amino acids.

> Some proteins (e.g. meat, fish, eggs, milk and soya beans) contain all the essential amino acids.
> Other proteins (e.g. peas, lentils, nuts and most beans) only contain some of the essential amino acids, so it's important to eat a wide variety of these foods. (This is particularly true for vegans — see p. 40.)

3) When you eat protein, your body breaks it down into amino acids and uses these to build new proteins — which your body then uses to make muscle, etc.

*There are **Three** Main Types of **Meat** Eaten in the UK...*

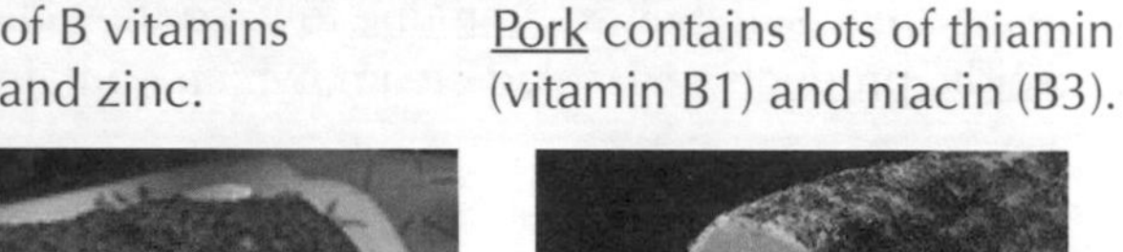

Beef and lamb have loads of B vitamins and minerals like iron and zinc.

Pork contains lots of thiamin (vitamin B1) and niacin (B3).

1) These are called red meats (though pork is sometimes classed as white meat). They're all great sources of protein, but too much can cause problems like heart disease.
2) Meat can be tenderised to make it, well... more tender. You have to partly break down the fibres in the meat. You can do this by bashing it with a mallet, marinating it in something acidic (see p. 38) or cooking it really slowly (this is what makes casseroles lovely and tender).
3) Meat can dry out during cooking. To avoid this you can seal the outside of the meat (by cooking it at a high temperature for the first couple of minutes) — this keeps the juices in.

*...and **Three** Main Types of **Fish***

Oily fish, e.g. herring, mackerel, salmon, tuna.

White fish, e.g. cod, haddock, plaice, skate.

Shellfish, e.g. crab, lobster, mussels.

Fish is very nutritious — it contains loads of vitamins, plus omega 3 oils, which are dead good for you.

Meat contains a lot of nutrients, but eating too much can lead to health problems

Meat's great, it's high in protein and it contains other nutrients as well — e.g. red meat has loads of iron, and liver has loads of vitamins. Sadly, some of the tastiest meat also has lots of saturated fat (see page 29). Shame.

Proteins — Meat, Poultry and Fish

Poultry contains lots of protein too, and it's also lower in saturated fat than many meats.
However, it's perfectly possible to get all the nutrients you need without eating any animal products at all...

There are **Three** Main Types of **Poultry** Eaten in the UK

Chicken

Turkey

Duck

These are white meats — though duck's often called red meat.

Poultry is a good source of protein and B vitamins and is fairly low in saturated fat (especially without the skin). But it can be contaminated with salmonella bacteria, which can make you seriously ill.

There are Now Loads of **Meat Replacements**

1) Vegetarians don't eat meat, so they need to get their protein and other nutrients from elsewhere. Beans, lentils and nuts are all good sources of protein, as are eggs (see next page).

Mixed beans

Green lentils

Mixed nuts

2) These days, there are lots of alternative proteins, such as:

Tofu — made from soya beans.
TVP (Textured Vegetable Protein) — also made from soya beans.
Quorn™ — made from a mushroom-like fungus and egg white.

3) These products can be prepared in lots of ways, sometimes to look like meat or chicken:

- TVP can be made into sausages, burgers and ready meals.
- Tofu is usually just stir-fried, but it can also be used in desserts.
- Quorn™ is more often used where you'd normally use chicken, and is available as chunks (e.g. for stir fries), mince (e.g. for chilli con carne) or fillets (e.g. to serve in sauces).

4) These meat replacements usually don't taste of much, so they're often flavoured. One way of doing this is by marinating them (soaking them in a mixture of things like oil, wine, vinegar and herbs) before cooking.

Meat replacements can be healthier than meat

As well as containing lots of protein, things like beans and lentils are low in fat and high in fibre, so they're a very healthy alternative to meat. They're also cheaper than meat, so they're widely used in poorer parts of the world.

Proteins — Eggs

Eggs are good sources of protein, vitamins and minerals, and they're also pretty useful when it comes to cooking.

Eggs Have Loads of Healthy Stuff

We mainly eat hens' eggs, but goose, duck and quail eggs are also popular with some people.

NUTRITIONAL CONTENT OF EGGS

- protein — about 13%
- fat (mainly saturated) — about 10%
- vitamins A, B2 and D
- minerals, including iodine

Eggs have Loads of Uses and Functions in Cooking

Functions

Binding, e.g. in burgers — coagulation sticks the ingredients together as they cook.

Thickening, e.g. in custard or quiche — egg white coagulates (becomes more solid) at 60°C and yolk at 70°C, so when it reaches these temperatures it sets and stays 'thickened'.

Coating or enrobing — eggs help dry ingredients like breadcrumbs to stick to food, e.g. chicken, as it's cooked.

Glazing, e.g. on bagels — brushing egg over bread gives it a glossy finish when it's cooked.

Aeration, e.g. in cakes — egg white traps air when it's beaten, because the protein stretches.

Emulsification, e.g. in salad dressings — oil and water mixed together form an emulsion (see page 46). But the emulsion usually separates after a while. Lecithin, found in egg yolks, keeps the emulsion stable (i.e. stops it separating again). That's why egg yolks are used in mayonnaise (see page 46).

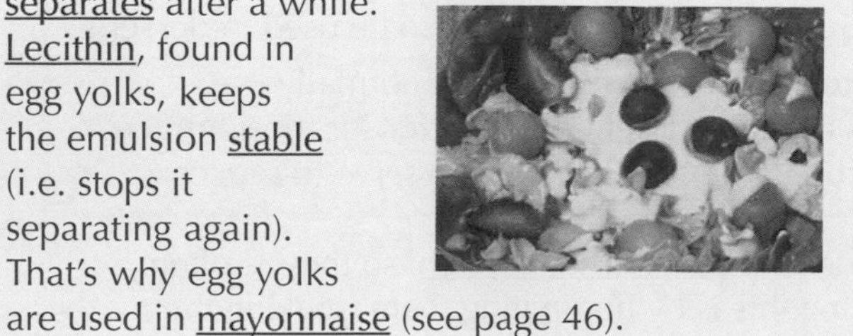

Make sure you know the different uses of eggs

Eggs are used left, right and centre in cooking so keep going over that list of uses and functions until you're absolutely sure you know them. Eggs aren't perfect though, they do have some problems — see the next page...

Proteins — Eggs

Eggs are great, but if they aren't cooked properly they can make you seriously ill.

Eggs May Contain **Salmonella**

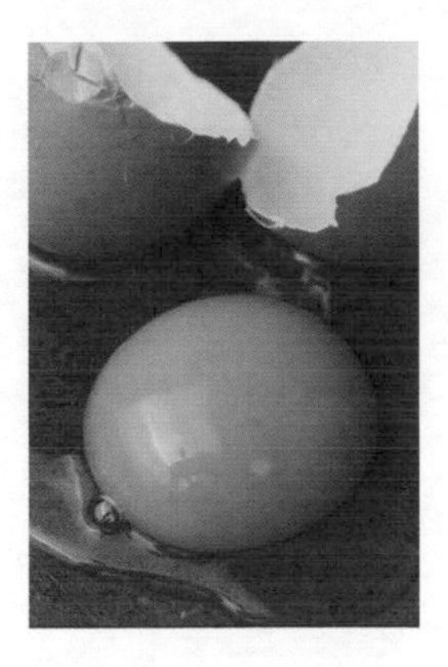

1) Raw eggs may contain the bacteria salmonella — which causes severe food poisoning. (You can also get it from chicken which hasn't been cooked properly.)
2) It's very important that eggs are cooked thoroughly so that all bacteria are destroyed.
3) You should be extra careful when cooking eggs to be eaten by pregnant women, babies and elderly or frail people.
4) Manufacturers often use dried or pasteurised egg to be on the safe side, like for mayonnaise.

EXAM TIP
If you're describing how to make a product that contains eggs, make sure you say how to ensure that they're prepared safely.

Be Careful with **How** You **Cook** Eggs

The way you cook an egg can make a lot of difference to how healthy it is.

BOILED AND POACHED EGGS

These are nice and healthy because they're cooked using no fat.

iStockphoto.com/Linda & Colin McKie

SCRAMBLED EGGS

These are healthy too — and if you do them in the microwave you don't need any fat.

FRIED EGGS

A lot of people like fried eggs — and these can absorb a lot of fat from the oil. It's best to use oils with unsaturated fat, and drain off as much of the oil as possible before eating them.

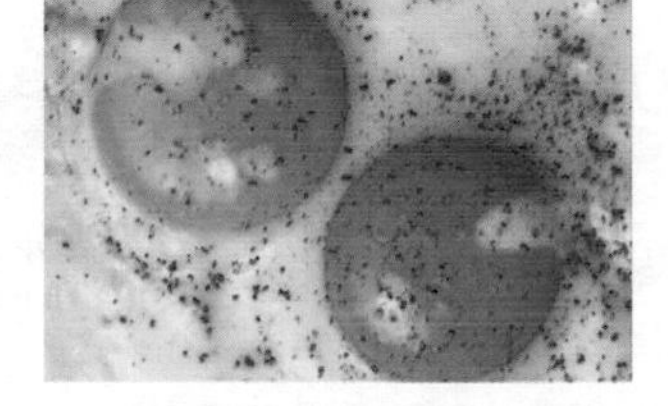

Bear in mind these issues when designing products

When you come up with a new food product you have to think carefully about safety (not making whoever eats it ill) and health. So it's well worth being aware of the problems of eggs and the different ways of cooking them.

Fats and Oils

Fats and oils are essential ingredients of many foods, and we also need a certain amount of them in our diet. Read on to find out more about the different types of fats and oils, and how they're used.

There are **Six Main Types** of Fats and Oils

1) Butter is made from churning cream.
2) Margarine is made from vegetable oils blended with a load of other stuff (which might include modified starch, water, emulsifiers...)
3) Lard is made from pig fat.
4) Suet is made from the fat which protects animals' vital organs.
5) Oils come from pressed seeds (e.g. rape seed, sunflower seed).
6) Low-fat spreads are emulsions of vegetable oils (usually hydrogenated — a process that makes them more solid) and water.

They're Used Loads in **Pastries** and **Biscuits**

Adding flavour — butter in shortbread and pastry makes them taste fantastic.

Shortening — rubbing fat into flour helps prevent gluten from being produced and makes pastry and biscuits 'short' — so they're a bit crumbly.

Adding colour — butter in pastry makes it golden yellow.

EXAM TIP
It's important to show the examiner that you understand both the good and bad points about fats and oils.

They're Used in **Other Types of Product** Too

- Cooking — deep frying (e.g. fish and chips) and shallow frying (e.g. eggs).
- Enriching — adding butter or cream to a sauce thickens it and makes it taste better.
- In emulsions — mixing together oil and water makes a thickish liquid, e.g. in vinaigrette — see page 46 for more about emulsions.

Learn the different types of fats and oils

Processed foods can contain loads of fat, so it's worth checking out the alternatives. We all need a certain amount of fat in our diets, although the type of fat is very important — have a look at the next page to find out why...

Fats and Oils

Fats and oils get a lot of bad press — everyone's always telling you to cut down on fat.
That's probably good advice, but don't go believing that they're all bad.

Fats have Some **Nutritional Value**

1) Fats are a concentrated source of energy.

2) Fats are a source of vitamins A, D, E and K.

3) Fats provide certain fatty acids which are essential to the structure and function of body cells.

4) The body needs a certain amount of fat to stay warm.

Unsaturated *Fats are* ***OK*** *but* ***Saturated*** *Fats are* ***Bad***

1) Saturated fats come mainly from animal sources (e.g. meat, butter, suet, dripping, lard) and are solid or semi-solid at room temperature. They're often associated with high amounts of cholesterol — see below.

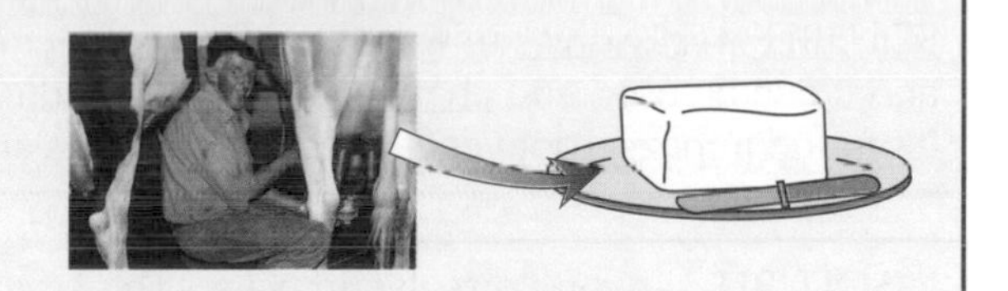

2) Unsaturated fats come mainly from vegetable sources and are usually liquid at room temperature. (Fats that are liquid at room temperature are called oils.) The main oils used in cooking are peanut, sunflower, corn, soya, rapeseed and olive oil.

Too Much Cholesterol *can be Dangerous*

Our bodies use fat to make cholesterol, which is an essential part of all cell membranes. It's also needed to make hormones.

But scientists think that high cholesterol levels can increase the risk of heart disease. Most people in the UK eat more saturated fat than the Government recommends — and could probably lower their risk of heart disease by cutting down.

See p. 39 for more on Government healthy eating guidelines.

We need some fat to survive...

...but too much can cause obesity, heart disease and all kinds of other health problems. Make sure you know the benefits of fat in the diet, the difference between saturated and unsaturated fat, and the problems caused by too much fat.

Fibre and Water

Fibre and water — more things you literally can't live without.

Fibre Isn't Digested by the Body

1) Fibre is a type of carbohydrate (see p. 20) — it's sometimes called roughage. It helps to keep your digestive system working properly and keeps food moving through it.
2) It's found in things like:

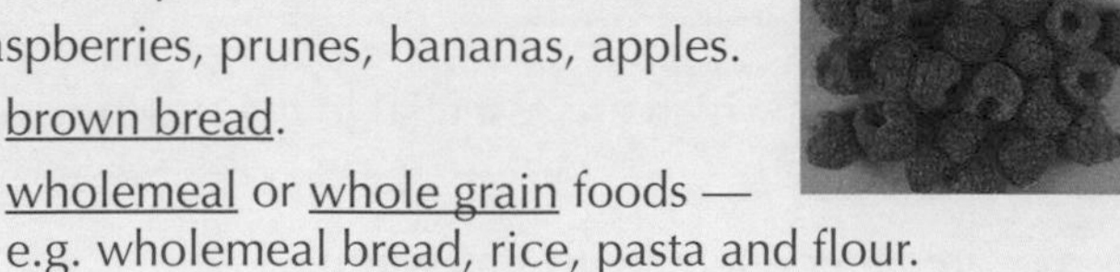

- vegetables — e.g. peas, beans, potatoes, broccoli, carrots.
- fruit and fruit juice — raspberries, prunes, bananas, apples.

- brown bread.
- wholemeal or whole grain foods — e.g. wholemeal bread, rice, pasta and flour.
- lentils, beans, seeds and nuts.

3) You need to eat lots of fibre to stay healthy. If you don't, it can lead to health problems such as: constipation (when you can't poo), bowel (intestine) and colon (part of the bowel) cancer, heart disease, high blood pressure and bowel diseases such as diverticulitis.
4) There are two types of fibre, and you need to eat both to have a healthy diet:

SOLUBLE — dissolves in water. It absorbs water as it passes through the body and it can be used and changed into other products by bacteria in the bowel. It's found in peas, root vegetables (like potatoes and carrots), fruit and oats.

INSOLUBLE — won't dissolve in water. This type of fibre passes through the body without changing at all. It's found in the skins of fruit and potatoes, whole grain and wholemeal foods, nuts and seeds.

You Can't Live Without Water

1) Around 75% of your body is water and all your body's chemical reactions and processes (e.g. digestion, excretion) take place using water.

2) You get water from drinks like water (obviously), fruit juice, tea, coffee, lemonade... Water's also found in food — vegetables and fruit contain quite a lot, and even things like meat and bread contain water.

3) If you don't have enough water from drinks and food you can become dehydrated — your body can't work properly. At first, you feel thirsty and produce less urine, and then you can get headaches or feel faint. Eventually, you can get cramp in your muscles and your blood pressure will drop. In extreme cases you can fall unconscious, go delirious and even die.

4) You should have about 2 litres of water a day — but if you're hot or exercising you need to drink more to get enough water into your system. People sometimes eat salty snacks with water or have isotonic drinks — this helps more water be absorbed by the body instead of it just passing straight through. They also help replace nutrients that are lost through dehydration.

Make sure you know what you get fibre and water from, and why you need them

There's a lot to take in on this page, so make sure you've got it all. Cover up the page and jot down why you need fibre, what foods contain it and what happens if you don't get enough of it. Then do the same for water.

Vitamins

Vitamins are essential for a healthy body — here are a few you should know about.

We need a Balance of Different Vitamins

Vitamin A

1) We get most of our vitamin A from retinol, which is found in liver, butter, fish oils and eggs.
2) We can also make it from carotene, which is found in orange or yellow fruit and veg and margarine.
3) Vitamin A is needed for good eyesight (especially night vision) and growth and functions of tissues.

Vitamin B Group (or Complex)

1) There are 8 different B vitamins (B1, B2, B3, B5, B6, B7, B9 and B12). Together they're known as Vitamin B Group (or Complex).
2) They're found in cereals, liver, kidney, peas, pulses, dairy produce, meat and fish.
3) B1, thiamin, helps the nervous system and the release of energy from carbohydrates. B2, riboflavin, helps with the release of energy and repair of tissues and B3, niacin, helps with the release of energy.
4) Folic acid is crucial for growth and important for women planning pregnancy, as low levels of folate at conception increase the risk of a baby having spina bifida.

Vitamin C (also known as Ascorbic Acid)

1) Vitamin C is found in citrus fruits (limes, oranges etc.), green veg, peppers and potatoes.
2) It's good for protecting the body from infection and allergies, helps in the absorption of calcium and iron from food, keeps blood vessels healthy and helps heal wounds.

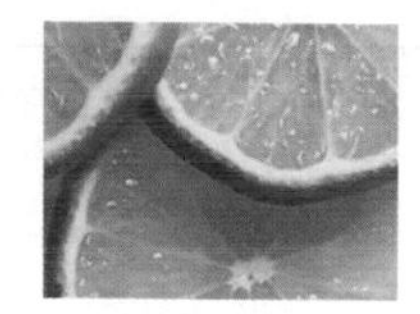

Vitamin D (also known as Calciferol)

1) Vitamin D is found in oily fish and eggs and is produced in the body when the skin is exposed to sunlight.
2) It's good because it helps the body absorb calcium.
3) A lack of it can lead to bone diseases like rickets and osteoporosis.

Learn the functions and sources of Vitamins A, B, C and D

The key things you need to remember from this page are what each vitamin is needed for, what happens if you don't get enough of it and which foods are good sources of each vitamin.

Vitamins and Minerals

The Government recommends that we eat 5 portions of fruit and veg per day. How we prepare those portions affects the amount of vitamins and minerals they contain.

We need a Balance of Different Minerals

Calcium

1) It's found in milk, tofu, salmon, green leafy vegetables, hard water and white bread.
2) It's needed for strong bones and teeth and healthy nerves and muscles.
3) Growing children need calcium every day for strong bones and teeth. Lack of calcium in youth can lead to problems in later life (e.g. osteoporosis).

Iron

1) Iron is found in dark green vegetables (e.g. spinach) and meat (especially liver and kidney).
2) It's needed to form part of the haemoglobin which gives blood cells their red colour. Lack of iron causes a deficiency disease called anaemia.

Sodium Chloride (Salt)

1) Salt is found in most foods, and some people add it to food as well.
2) It's needed to regulate the water content in the body, but too much salt's bad for you — it can lead to high blood pressure and heart disease.

Phosphorus

1) Phosphorus is found in foods like meat, fish, dairy products, nuts, beans and cereals.
2) It's needed for healthy bones and teeth. A lack of phosphorus can lead to weak muscles and painful bones.

Fruit and Vegetables are Healthy

In a normal healthy diet, fruit and vegetables give you:

- The majority of your vitamin C intake (about 90%)
- Dietary fibre
- Vitamins A (from carotene) and B
- Iron and calcium
- Loads of water
- Not much fat (except avocados)
- Small amounts of protein.

Prepare Fruit and Veg Carefully to Keep the Good Stuff

Nutrients and flavour can easily be lost or spoilt through overcooking and poor storage.

1) Microbes (germs) in the air can make fruit and vegetables go rotten. Store them in a cool, dark place (e.g. a fridge or larder, depending on the food).
2) Prepare fruit and vegetables just before you need them — vitamin C, in particular, starts to go once the fruit and vegetables are picked, stored, cut or peeled.
3) Don't chop fruit and vegetables into small pieces — it exposes more of the surface and more nutrients are lost when they're cooked.
4) Don't leave vegetables to stand in water — vitamins B and C dissolve into the water.
5) Most of the nutrients and the fibre are found just below the skin of fruit and vegetables, so peel very thinly or use them cleaned and unpeeled if possible (like jacket potatoes).
6) Fruit and vegetables should be cooked as quickly as possible in a small amount of water. Steaming or microwaving them are the best ways to keep the nutrients.

Bananas give off a gas which makes other fruit and vegetables ripen quickly and spoil, so they need to be stored separately.

The way you store and prepare food affects how healthy it is

A lot of this stuff on how to prepare food may seem like common sense, but you need to know the theory behind the practice if you're going to impress the examiner. Close the book and scribble down as much of this page as you can.

Warm-Up and Worked Exam Questions

Phew, there was a lot to take in there. Check you understand it all by having a go at these questions.

Warm-up Questions

1) Explain why sugar is used in the following foods:
   a) jam b) bread c) biscuits
2) Mary is going to cook steak for dinner. Before she fries it, she bashes it with a mallet. Explain why she does this.
3) Describe three ways you can reduce the risk of getting salmonella poisoning.
4) Explain why a diet containing no fat would be unhealthy.
5) Sam is 12 years old and has had several broken bones recently. He also often suffers from colds. What kind of foods would you recommend he eats more of? Explain your answer.

Worked Exam Questions

1 A pie company wants to improve the pastry in their products.
Sensory tests have shown their pastry is too crumbly.

(a) Suggest how they could modify their recipe to improve the pastry texture.
Give a detailed reason for your answer.

The question asks for a detailed reason, so make sure you give one.

They could add less fat to their pastry mixture. This would mean more gluten could be produced when mixing the flour and other ingredients so the pastry would stick together better.

(3 marks)

(b) The company wants the top of their pie to be golden-yellow in colour.
Name one type of fat that could be added to the pastry mixture to get this appearance.

Butter.

(1 mark)

2 Vegetarians can find it hard to get enough iron in their diet.

(a) Name one vegetarian food rich in iron.

Dark green vegetables, e.g. spinach. *It's always a good idea to give an example.*

(1 mark)

(b) Briefly explain why iron is needed as part of a healthy diet.

It's needed to form haemoglobin/blood.

(1 mark)

3 A restaurant serves carrots with every main meal. They use fresh carrots, which are peeled and sliced before being cooked in boiling water.

(a) Suggest three ways to minimise the nutrient loss from the carrots as they are cooked.

Cook the carrots as quickly as possible in a small amount of water. Peel the carrots very thinly. Steam or microwave the carrots to keep in more of the nutrients.

(3 marks)

(b) What effect would freezing the carrots before cooking have on their nutritional value?

No effect.

(1 mark)

Exam Questions

1 A basic bread dough is made from flour, water, yeast and salt.

(a) Name the type of flour most suitable to use in bread-making.

....

(1 mark)

(b) Name the protein in bread dough that makes it elastic.

....

(1 mark)

(c) Describe the function of the yeast in bread dough.

....

(1 mark)

2 A basic swiss roll is made from sponge cake, jam and cream. Describe how a basic swiss roll product could be adapted to meet the needs of a consumer on a calorie-controlled diet. You may use annotated sketches in your answer.

(3 marks)

3 Eggs are a high protein food.

(a) Explain why protein is essential in a person's diet.

....

....

....

(3 marks)

(b) Name two other nutrients found in eggs.

1. ....

2. ....

(2 marks)

Additives

There are ways of improving food products so they're nicer than ever.

Additives are Really Useful Substances Added to Food

1) An additive is something that's added to a food product to improve its properties.
2) Additives have loads of different uses — from improving taste to extending shelf life.
3) Some additives occur naturally and some are made artificially. Customers tend to prefer the idea of natural additives, so manufacturers try to use these where possible.
4) All additives must pass a safety test before they can be used in food. When an additive passes it gets an E number, meaning it can be used throughout the European Union, e.g. caramel colouring is E150a.

1 Preservatives

Preservatives are additives that prevent bacteria from growing — so the food lasts for longer.

EXAMPLES

- Vinegar is used to pickle foods like onions and eggs.
- Using concentrated lemon juice keep salads fresh.
- Salt is used to cure meat, e.g. ham, bacon.
- The sugar in jam preserves it.
- All of these are natural preservatives.

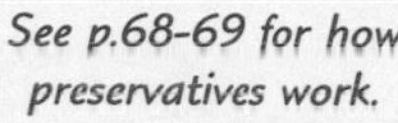
See p.68-69 for how preservatives work.

2 Colourings

1) Colourings make food look more attractive and more appealing to eat.
2) They can be used to add colour to something colourless, or to return food to its natural colour if it's lost during processing.

EXAMPLES

- Caramel is a natural food colouring — it's used to make products darkish brown, e.g. soft drinks like cola.
- Tartrazine is an artificial food colouring — it's used to make products a yellow colour, e.g. custard powder, syrups, sweets.

3 Flavourings

Flavourings improve the taste or the aroma (smell) of a product.

EXAMPLES

- Herbs and spices are natural flavourings — they improve the taste, e.g. adding basil makes tomato-flavoured pasta sauces more tasty, and chillies add spice to a range of foods.
- Vanilla flavouring can be natural (from vanilla pods) or artificial (vanillin solution), e.g. vanilla essence. It's used in lots of cakes and desserts, e.g. vanilla-flavoured ice-cream.
- Artificial sweetening agents, e.g. saccharin, are used in some desserts to...erm, add sweetness.
- Monosodium glutamate (MSG) is a natural flavour enhancer — it boosts the existing flavour of a product and gives it a savoury taste. MSG is added to processed foods like sauces, soups and crisps.

Additives make food look better, taste better, or last for longer

Unless you grow everything you eat (unlikely), you'll probably be eating a fair few additives — they're extremely useful. So make sure you know why they're useful, and learn some examples too.

Additives

Additives aren't just used to preserve, colour or flavour foods — they have other uses too. It's not all plain sailing though, additives can sometimes cause problems, and you need to be aware of what these problems are.

4 Emulsifiers

Emulsifiers are used to keep food products stable — they stop oily and watery liquids separating.

EXAMPLE

Lecithin is a natural emulsifier found in egg yolks — it's used in products like mayonnaise and margarine.

See p. 46 for more on emulsifiers.

5 Setting Agents

Setting agents are used to thicken products, so that they set as a gel (see p. 46).

EXAMPLE

Gelatine is a natural gelling agent that's extracted from animals — it's used in desserts like mousses and jellies.

6 Raising Agents

Raising agents are used in dough and cake mixtures to aerate them. They release bubbles of gas, which expand when heated to make the mixture rise.

Fermentation is when yeast breaks down the sugars in the dough releasing carbon dioxide and alcohol (which evaporates).

EXAMPLES

- Yeast is a biological raising agent used in bread dough — yeast are microorganisms that cause fermentation, producing carbon dioxide.
- Baking powder and bicarbonate of soda are chemical raising agents. They break down when heated, producing carbon dioxide which makes cakes rise.

But Food Additives have **Disadvantages** Too

1) Some people, especially kids, are allergic to certain additives.

2) Some additives, like sugar or salt, if used in large amounts can be bad for our health.

3) They can disguise poor quality ingredients, e.g. processed meat products may not contain much meat but they can be made to taste good by using additives.

4) Although additives go through safety tests, no-one really knows the long-term health effects yet. Some people think eating additives could be linked to behavioural problems, e.g. studies are looking at whether a colouring additive called sunset yellow is linked to hyperactive behaviour in children.

Additives are useful, but they also cause problems

Generally, highly processed foods are likely to contain lots of additives — and this can be as a substitute for high quality ingredients. Most additives are fine in moderation, but some are known to cause health problems.

Acids and Alkalis

'Acids and alkalis' may sound more like chemistry than food technology, but unfortunately you've got to learn about them. The good news is you don't have to know the chemistry behind them in any great detail.

Acids and Alkalis Change the Properties of Foods

1) Food and ingredients can be acidic, neutral or alkaline.

2) Acids and alkalis have a big effect on the flavour, texture and appearance of foods.

3) So you need to know the pH of your ingredients to know how your final product will turn out.

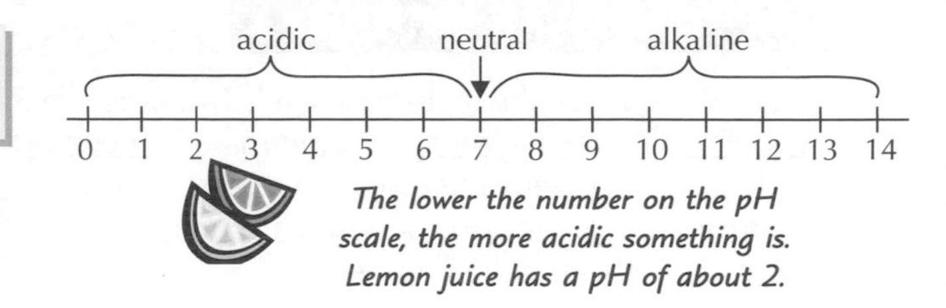

The lower the number on the pH scale, the more acidic something is. Lemon juice has a pH of about 2.

Alkalis have a Couple of Uses...

Foods like bicarbonate of soda and cornflour are alkaline. They have a fairly unpleasant, bitter taste.

1 Bicarbonate of soda acts as a raising agent

There's more about raising agents on p. 36.

Bicarbonate of soda breaks down to produce carbon dioxide when it's heated. The carbon dioxide bubbles expand and make mixtures rise.

You need to use bicarbonate of soda together with a strong flavour, to mask the unpleasant taste it leaves — so it's used in things like gingerbread and chocolate cake.

2 Cornflour gives a thicker texture

Cornflour is used in lots of products to thicken them. For example, you add cornflour to thicken the filling in good old lemon meringue pie.

The pH of food can affect how it looks and tastes

In the exam, you could be asked a straightforward question like 'Name an alkali that acts as a raising agent', or you could be asked to use your knowledge to solve a cookery problem — so learn this page well.

Acids and Alkalis

Acids have even more uses than alkalis — here are some common ones you should know.

Acids have Loads of *Different Uses...*

Acidic foods include citrus fruits, lemon juice (citric acid), vinegar (acetic acid) and vitamin C (ascorbic acid). They have a sharp, sour taste.

1 Vinegar gives a softer texture

Acid can change the texture of foods by partly breaking down proteins. If you add vinegar when making meringue (e.g. for a dessert like pavlova) it gives the meringue a softer, chewier texture. Vinegar is also used in marinades to soften the texture of meat — this is called tenderising.

2 Lemon juice prevents enzymic browning

- When you slice fruits (e.g. pears), the inside surfaces of the fruit react with oxygen in the air — the reaction is called enzymic browning and it turns the fruit brown.
- But if you dip your slices of fruit into lemon juice straight away, the acidic conditions stop enzymic browning. The colour of the fruit is retained, which is really useful, e.g. for the nice appearance of fruit salads.

3 Lactic acid fermentation produces yogurts

Milk turns sour when the bacteria it contains break down sugars in the milk into lactic acid. In yogurt-making, lactic acid acts on the proteins in milk to thicken it. The lactic acid gives yogurt its slightly sour taste.

4 Acids add flavour

Acids are added to give a sharp flavour, e.g. vinegar is often used in salad dressings.

5 Acids help preserve foods

The acidic conditions help to preserve foods because bacteria can't grow (see p. 69).

EXAM TIP
Make sure you know some specific examples of what acids and alkalis are used for in food production.

Acids taste sour, alkalis taste bitter

Hmm, I don't know about you but adding vinegar to a meringue just doesn't seem right somehow. Still, as long as it tastes good I guess it's OK. Now, close the book and write down the five uses of acids described here.

Healthy Eating

The Government issues guidelines on healthy eating. Food labels can help you work out if you're following the guidelines.

Use the **Eatwell Plate** to Check your **Diet** is **Right**

The eatwell plate is an easy way of showing how much or little of each food group you should eat:

LOTS of fruit and vegetables.

The Government recommends you eat at least 5 portions of fruit and veg every day to get enough vitamins and minerals. (one portion = 80 g)

LOTS of starchy foods, like pasta, rice, bread and potatoes.

SOME non-dairy sources of protein, e.g. meat, fish, eggs, beans.

SOME dairy foods, e.g. milk, cheese, yogurt.

SMALL AMOUNTS of fatty and sugary foods.

You Need Certain Amounts of Each **Nutrient**

Many food labels show you how much of various nutrients the product contains.
They often refer to Guideline Daily Amounts (GDAs) or Recommended Daily Amounts (RDAs) — this is how much of each nutrient and how much energy an average adult needs each day.

Labels usually show how much protein, carbohydrate, fat and dietary fibre the product contains.

The amount of each nutrient is sometimes shown as a percentage of the GDA as well.

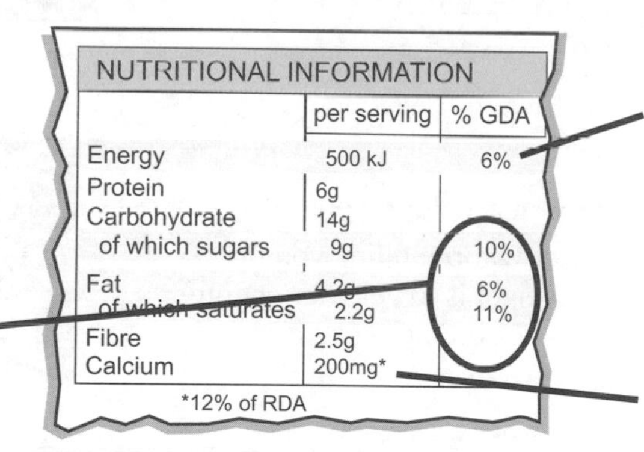

NUTRITIONAL INFORMATION

	per serving	% GDA
Energy	500 kJ	6%
Protein	6g	
Carbohydrate	14g	
of which sugars	9g	10%
Fat	4.2g	6%
of which saturates	2.2g	11%
Fibre	2.5g	
Calcium	200mg*	

*12% of RDA

One serving of this product contains 6% of the daily energy requirement.

RDAs are used for vitamins and minerals. One serving of this product contains 12% of the recommended daily calcium intake.

A diet with not enough or too much of certain nutrients can cause health problems:

Nutrient deficiencies

1) Not getting enough vitamins or minerals can lead to many health problems, see p. 31.
2) Not eating enough protein leads to restricted growth in children, and also leads to muscle wastage.

Nutrient excesses

1) Eating too much fat can make you overweight or obese, which can lead to heart problems and cancer.
2) Eating too much sugar can lead to obesity, Type 2 diabetes and tooth decay.
3) Too much salt can increase blood pressure, meaning a bigger risk of heart disease and stroke.

Practise reading food labels to make sure you understand them

You can check how healthy your diet is by keeping a food diary of everything you eat, and comparing this to the guidelines. The important thing is to get a balanced diet over a day or so. And to revise.

Healthy Eating

Some people can't or won't eat some foods. When you're designing a product, you need to think about the ingredients you'll be using and whether they'll be suitable for different groups of people.

Different Groups have Different Dietary Needs

Vegetarian and Vegan

1) Vegetarians don't eat meat or fish.
2) Vegans don't eat any animal products, e.g. they won't eat milk or cheese.
3) Vegetarians and vegans need to get protein and vitamins from foods like nuts, beans and lentils or from meat replacements (see p. 25).

Coeliac

People with coeliac disease can't eat a protein called gluten, found in wheat, rye and barley.

1) So coeliacs can't eat normal bread or pasta.
2) They have to get starch and fibre from other foods, e.g. rice and potatoes, or from gluten-free alternatives, e.g. gluten-free bread.

Lactose Intolerant

People who are lactose intolerant can't digest lactose, which is a sugar found in milk.

1) So they need to avoid cow's milk and many dairy products, and any products that lactose is added to, e.g. peanut butter.
2) Dairy products are an important source of calcium for most people. Lactose intolerant people have to get calcium from foods like green leafy vegetables, salmon and white bread.

Calorie-Controlled

1) People who are overweight need to eat a calorie-controlled diet.
2) Fat and sugar provide a lot of calories without filling you up. The recommended method to lose weight is to get your energy from starchy foods instead (and to take a bit more exercise).

Diabetic

1) Diabetics can't control their blood sugar levels.
2) They need to avoid eating very sugary foods.
3) Diabetics are advised to eat plenty of starchy foods, which release energy slowly, to avoid having very high or very low blood sugar levels.

Nut Allergy

1) People with nut allergies need to avoid products that contain nuts or have traces of nuts.
2) If you're designing a product containing nuts, consider putting a warning label on the packaging.

Make sure you know the needs of each dietary group on this page

Some people choose a specific diet for health, moral or religious reasons, whilst other people can't eat certain foods for medical reasons. Close the book and jot down each group and their dietary requirements.

New Technology

Get scientists involved in food technology and there's no end to what you can do...

There are **New Production Methods**, **Foods** and **Packaging**

1) Producers use new technologies to meet consumer demands, e.g. for food that lasts longer.
2) New methods of producing food all year round include using huge greenhouses called biodomes.
3) New ingredients and foods include meat substitutes like TVP and tofu, modified starch (see p. 23), genetically modified foods and functional foods (see next page).
4) New packaging technology includes breathable packaging for fruits, and packaging with more protection against moisture and bacteria — see page 84 for more.

Genetically Modified Foods have **Altered Genes**

1) A genetically modified (GM) food is one that's had its genes altered to give it useful characteristics. GM plants are produced by inserting a desirable gene from another plant, an animal or a bacteria into the plant you want to improve. You plant modified seeds and up comes your GM crop.
2) For example, you can get GM maize that's pest-resistant. The farmer gets a bigger yield of maize because less of the crop is eaten or damaged by pests.

GM foods have advantages...

1) Crops can be made to grow quickly.
2) Producers can get a higher yield of crops for the same amount of seed and fertiliser.
3) This makes food cheaper to produce so it's also cheaper for the consumer to buy.
4) Crops can be altered to have a longer shelf life — so less food is wasted.
5) Crops can be made to ripen earlier than normal, so fresh foods can be available for consumers earlier in the year.

...and disadvantages

1) GM foods haven't been around for long — so their long-term health effects aren't known.
2) There are concerns that modified genes could get out into the wider environment and cause problems, e.g. the weedkiller resistance gene could be transferred to a weed, making it a 'superweed'.
3) GM producers can't sell their food everywhere — the European Union (EU) restricts the import of some GM foods.

Consumers and the **EU** have **Safety Concerns**

Some people believe that we shouldn't mess about with genes because it's not natural.
To help consumers make an informed choice, the European Union (EU) has rules:

1) All GM foods must undergo strict safety assessments and they can only be sold if they're found to have no health risks.
2) All foods that are GM or contain more than 1% GM ingredients must be clearly labelled.

Using new technology has many advantages in food production

New technology can help food producers and consumers... but not everyone's convinced that the benefits outweigh the risks. Make sure you know the pros and cons of GM foods and what the EU says.

New Technology

As well as genetically modifying foods to increase food production, manufacturers can also modify and add things to foods to make them more nutritious — these foods are called functional foods.

Functional Foods have Added Health Benefits

A functional food is one that's has been artificially modified to provide a particular health benefit, on top of its normal nutritional value. For example:

- Some fruit juices have calcium (which is important for healthy bones) added to them.
- Eggs containing high levels of the fatty acid omega-3 (which may reduce your risk of heart disease and cancer) can be produced by feeding hens a diet rich in omega-3.
- Some functional foods are made by genetic modification, e.g. Golden Rice is rice that's been genetically modified to contain carotene (which provides vitamin A, important for good eyesight).

When extra nutrients are added to foods it's called fortification.

They solve some problems...

1) Functional foods are an easy way of providing better nutrition for people who have a poor diet.
2) People who can't eat (or don't like) certain foods can get the 'missing' nutrients from functional foods.
3) Foods like Golden Rice could help solve some health problems caused by malnutrition in poor countries.

...but not all

1) People still need to eat a varied diet and exercise to be healthy — they can't rely on just eating functional foods.
2) Functional foods don't always provide all of the nutrient you'd need. E.g. it might be difficult for people to eat loads of Golden Rice every day to get enough vitamin A.
3) They don't tackle the actual causes of malnutrition in poor countries.

But Some Consumers are Worried

Many consumers are concerned about what's in food and they don't know whether to believe everything manufacturers tell them.

1) If manufacturers make health claims for their foods, e.g. 'helps to maintain a healthy heart', then the relevant ingredients must be clearly labelled and the nutritional information must be on the label (see page 82 for more on labelling laws).
2) Health claims mustn't mislead consumers, and must be backed up by scientific evidence.

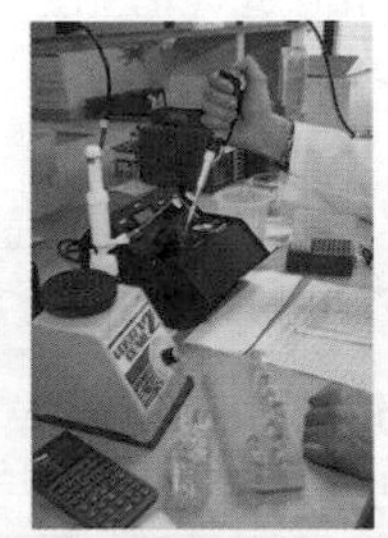

Scientists have to test food rigorously to check that it has the health benefits it claims to have.

Learn the definition, and the pros and cons of functional foods

Functional foods are potentially really useful in making sure that people get enough of the right nutrients. Make sure you know why functional foods can be useful, a couple of examples and why people have reservations about them.

Warm-Up and Worked Exam Questions

You're nearly at the end of this section, so you should know all there is to know about the properties of food. Make sure you do by having a go at these warm-up questions, before moving on to the exam-style questions.

Warm-up Questions

1) Sophie's made some sweet chilli sauce, but she isn't convinced about the taste.
   a) What could she add to make the sauce have a more savoury taste?
   b) What could she add to make it a bit sweeter?
2) What kind of flavour do strongly acidic foods have?
3) Susie is hosting a dinner party and plans to serve beef lasagne.
   a) One of her friends is vegetarian. Suggest how Susie could make her a suitable lasagne.
   b) One of her friends is a coeliac. Why might this be a problem for Susie and how could she alter her lasagne to make it suitable?
4) What is a genetically modified (GM) food?

Worked Exam Questions

1 Some additives are natural and others are artificial.

(a) Suggest one reason why a manufacturer might prefer to use natural additives in their products rather than artificial ones.

Because customers don't tend to like the idea of artificial additives, so are more likely to buy products containing natural additives.

Remember that manufacturers are mainly concerned with selling products, so they'll avoid anything that might reduce sales.

(1 mark)

(b) A caramel and vanilla slice contains the following ingredients:
Flour, eggs, milk, butter, salt, jam, sugar, caramel, saccharin, vanilla essence.

(i) Name two natural additives from the ingredients list.

Salt, sugar

You could also say caramel.

(2 marks)

(ii) Name two artificial additives from the ingredients list.

Saccharin, vanilla essence.

(2 marks)

(c) Suggest two reasons why the additives you named in part (b) might be used in the slice.

To add flavour and to add colour.

(2 marks)

2 A company is developing a fruit pavlova product.

(a) During the product development, vinegar was added to the meringue mixture. Briefly describe and explain the effect this would have had on the meringue.

It will give the meringue a softer, chewier texture because vinegar partly breaks down the proteins in the mixture, which changes the texture.

Don't forget to explain your answer. The explanation's only worth 1 mark, so you don't need to go into loads of detail.

(2 marks)

(b) Name an alkali that could be added to give the meringue a thicker texture.

Cornflour.

(1 mark)

Exam Questions

1 Look at the extract from a food diary shown below.

Tuesday	Breakfast:	2 pieces of toast with margarine 1 glass of orange juice
	Lunch:	75 g of cheese, melted on two pieces of toast
	Dinner:	steak pie and a portion of chips
	Additional snacks:	1 chocolate bar (30 g) 1 packet of crisps (35 g)

(a) Suggest two ways in which the person keeping the diary could improve their diet.

....

....

(2 marks)

(b) The person keeping the food diary is diabetic.
Name one food listed that they should avoid eating.

....

(1 mark)

2 A can of soda lists the additive sorbitol, which has the E number E420, among its ingredients. Describe what the term 'E number' means.

....

....

(2 marks)

3 New technologies have allowed manufacturers to develop functional foods.

(a) Explain what is meant by the term 'functional food'.

....

....

(2 marks)

(b) Give two examples of functional foods.

1. ....

2. ....

(2 marks)

(c) Describe how functional foods can improve people's health.

....

....

....

....

(3 marks)

Revision Summary for Section Two

I know you're eager to rush off for a cuppa and a slice of cake to celebrate getting to the end of this section, but before you do, have a go at these revision summary questions — they're here to help you.

1) Why is starch a good thing to eat a few hours before running a marathon?
2) What happens if you eat loads of carbohydrates but don't use the energy?
3) Name four types of sugar.
4) Give two reasons why people might use sugar substitutes.
5) a) Briefly describe what gelatinisation means.
   b) Over what temperature range does gelatinisation occur?
6) Name two ways that modified starches can be useful in food preparation.
7) Why does kneading bread dough help it to rise?
8) What do we need proteins for?
9) List three examples of foods that contain all the essential amino acids.
10) Name two vitamins or minerals that are contained in each of the following:
    a) beef b) pork
11) Give one advantage and one disadvantage of eating white meats.
12) a) What does "marinate" mean?
    b) Why are alternative protein foods often marinated?
13) Explain how eggs can be used in each of the following ways, and give an example product or food for each:
    a) aeration b) binding c) thickening
14) Why are fried eggs less healthy than boiled eggs?
15) List six types of fats and oils, and describe briefly how each is made.
16) What's the difference between saturated and unsaturated fats?
17) Name four foods that are a good source of fibre.
18) What are the two different types of fibre?
19) What can happen if you get dehydrated?
20) What foods do we get vitamin A from? Why do we need it?
21) List four types of B vitamin. For each one, state why it's useful.
22) a) List five good sources of calcium.
    b) Why do our bodies need plenty of calcium?
23) Why is iron good for us? Name two good sources of it.
24) a) List three ways in which nutrients can be lost from fruit and vegetables during preparation.
    b) For each one, describe how you can prevent this from happening.
25) What is an additive?
26) Explain why manufacturers add the following types of additive to food products:
    a) preservatives b) colourings c) setting agents d) emulsifiers
27) Describe a common function of bicarbonate of soda.
28) Roberto is about to prepare a fruit salad. It won't be served till later, but he wants it to look really fresh.
    a) Name an acid or an alkali that you would recommend he uses in the fruit salad.
    b) Explain how your recommended ingredient would help.
29) List the five major food groups on the eatwell plate.
30) Explain how eating too little protein can be bad for you.
31) What nutrient might lactose intolerant people not get enough of? What can they eat to get this nutrient?
32) a) Describe three benefits that GM foods can bring.
    b) Give two potential disadvantages of GM foods.
33) How does the European Union (EU) help consumers make informed choices about GM food?
34) A label on a carton of orange juice says it 'helps to maintain healthy bones'.
    What does the manufacturer need to put on the label to back up this claim?

Combining Ingredients

How a product turns out depends on the ingredients you use and how they react with each other.

Solution — a Solid is Dissolved in a Liquid

1) A solution forms when solid ingredients completely dissolve in a liquid, e.g. sugar dissolves in water to make a solution.
2) You can't tell the separate ingredients apart once they've formed a solution.

sugar + water → a solution

Suspension — a Solid is Held in a Liquid

1) A suspension forms when solid ingredients are added to a liquid but don't dissolve, e.g. flour mixed with water forms a cloudy suspension.
2) If you don't stir the mixture, the solid usually sinks to the bottom.
3) If you stir and heat up a starchy suspension, it'll thicken into a sauce (see page 22).

Emulsion — Oil and Watery Stuff Mixed Together

1) An emulsion is formed when oily and watery liquids are mixed together and the droplets of one spread out through the other — they usually separate unless you keep shaking or stirring them.
2) Emulsions need an emulsifier to stop the oil separating from the liquid.
3) There's a natural emulsifier in egg yolks — it's called lecithin.
4) Egg yolks are used as emulsifiers in foods like margarine to hold together the oil and the liquid.
5) Many types of salad dressings are emulsions, e.g:

- Mayonnaise is a stable emulsion of egg yolk, oil and vinegar (often with other flavourings).
- Vinaigrette is an emulsion of oil and vinegar. It's unstable — it separates if you don't shake it.

Gel — a Small Amount of Solid Sets a Lot of Liquid

1) A gel is a thick solution, in between a liquid and a solid. (Well, not exactly... but you don't need to know the full chemistry explanation.)

2) Some cold desserts are gels, e.g. jellies, mousses, cheesecakes.
3) Only a small amount of solid ingredient is needed to set a lot of liquid, e.g. a small amount of gelatine can set a lot of water to form jelly.
4) There's a natural gelling agent in some fruits — it's called pectin. Pectin is released from these fruits during cooking and it helps foods like jams to set.

Ingredients can behave in different ways when mixed together

The key to making loads of food products is to know what happens when you combine different ingredients together. Make sure you learn an example of a solution, a suspension, an emulsion and a gel.

Combining Ingredients

The ingredients you choose to combine affect the flavour and nutritional value of the final product.

You can Change the Taste of your Food Product...

Food products are made to meet specific criteria (see page 9). You can make your food product exactly how you want it (the right colour, texture, taste, etc.) by doing these things:

You can change the ingredients

To make a cake look darker you could swap white flour for wholemeal flour. Or to give it a different flavour you could vanilla essence or lemon zest.

You can change the proportion of ingredients

To make a cake taste nuttier you could add more nuts. Or to make it more moist you could add more eggs to the mixture.

You can change the way you combine the ingredients

To make a cake more light and fluffy you could beat the egg whites for longer. If your cake's a bit too heavy you could try folding the flour into the mixture instead of beating it.

You need to accurately measure and record any changes you make to your product — so you can make it exactly the same next time. This is really important when making your product on a large scale — any measuring errors could be expensive and waste time if you make a rubbish batch or run out of ingredients. Use things like measuring spoons, measuring jugs and electronic scales to measure everything accurately.

...and you can Change its Nutritional Value

If you want to change the nutritional value of your food product (to meet specific criteria) then you need to change the ingredients or the proportions of ingredients.

1) Low-sugar cakes and biscuits — you could experiment using less sugar to make a cake or biscuits. You might have to swap sugar for another sweet ingredient, e.g. honey or fruit. But you'll have to experiment to get the proportions and the cooking time right.
2) Low-fat pastry — you could try using less fat in pastry, e.g. use less butter, but you'd have to make sure the texture of the pastry was still suitable for the product you were making.
3) Gluten-free bread — you could make bread using gluten-free flour, but the texture would be very heavy without any gluten to make the dough elastic. To avoid this, you could add xanthan gum to the dough — this makes it 'stickier' and gives the final product a much better texture.
4) Low-salt sauces — you could make a sauce with very little salt — but you might have to give it more flavour some other way, e.g. use less salt in tomato sauce but add some basil instead.

Getting the right combination can make your recipe a great success

Getting the flavour and nutritional value of your products right requires trial and error (see pages 11-12), e.g. you may be able to tell that your pastry needs more butter but you won't know exactly how much until you give it a go.

Standard Food Components

In industry, manufacturers don't always make their products from scratch...

Standard Food Components are Ready-Made Parts

1) Manufacturers can buy in food parts that have already been made by other manufacturers, e.g. pizza bases, fillings. These ingredients are called standard food components. They're pre-manufactured components.
2) Standard food components are really useful — they cut out a lot of time and work.
3) It's not just manufacturers — catering businesses and people cooking at home use standard food components too (see next page).
4) Standard components are processed foods — they've been prepared, treated or altered in some way. Highly processed foods often contain lots of added sugar and salt and can be bad for your health. Not all processed foods are bad for you though — things like milk and fruit juice are processed but are still good for you.

Using Standard Food Components has Advantages...

1) It saves time — you don't have to bother preparing the basic ingredients. This can improve your quality of life, as you get to spend more time with your family.
2) It saves money — manufacturers can often buy standard components frozen, in bulk, which is more cost-effective than buying fresh ingredients separately and making the components themselves.
3) Less machinery and less specialist equipment is needed, which also saves money.
4) Fewer specialist skills are needed by staff because the standard components are ready to use.

Some pizza delivery shops use ready made pizza bases so they can get pizza to you faster.

5) Food preparation is safer and more hygienic — especially if high-risk products like chicken, eggs or soiled vegetables are stored and prepared somewhere else.
6) The product is more likely to be consistent — standard food components are quality-controlled so they all have the same flavour, texture, weight, shape, colour, etc.

...and Disadvantages

1) You can't pick and choose exactly what you want, e.g. you can't get the ready-made pastry made a tiny bit sweeter.
2) It's not always reliable — late deliveries from the supplier will hold up the production line.
3) The product may not be as tasty as one made with fresh ingredients.
4) Extra space might be needed to store the standard food components if you've bought them in bulk.
5) There may be extra packaging and transport involved, so it might be bad for the environment.

Standard food components make life easier

They're really useful for manufacturers because they make the production process quicker and easier, and help with product consistency. However, it means they have to rely on other manufacturers, which can cause problems.

Standard Food Components

Now you know what <u>standard food components</u> are, it's time to learn some <u>examples</u>.

Ready-made **Pastry**, **Pizza Bases** and **Cake Mixes** Save **Time**

1) Standard food components include things like <u>pizza bases</u>, <u>chilled and frozen pastry</u>, <u>cake mixes</u> and <u>bread mixes</u>.
2) It's <u>much quicker</u> to start off with a ready-made component than to make it yourself.
3) You can <u>adapt</u> it as you need, e.g. you can put your own fillings into pies but start off with ready-made <u>pastry</u>.

Ready-made **Fillings** and **Sauces** have a **Long Shelf-Life**

1) <u>Ready-made fillings</u> and sauces save you the bother of having to prepare all the separate ingredients.
2) They also have a <u>longer shelf life</u> than fresh products and can be <u>used at any time of year</u>, not just when the fresh products are in season.
3) For example, you could make a <u>blackberry pie</u> in <u>January</u> if you used pie filling.

Ready-to-roll **Icing** and **Marzipan** make **Cake Decoration Easier**

1) <u>Ready-to-roll icing</u> and <u>marzipan</u> are easy ways to <u>decorate</u> products without having to prepare everything yourself.
2) For example, a cake business might order in ready-made <u>marzipan</u> in various colours, but <u>cut and shape it themselves</u> to make their decorations.

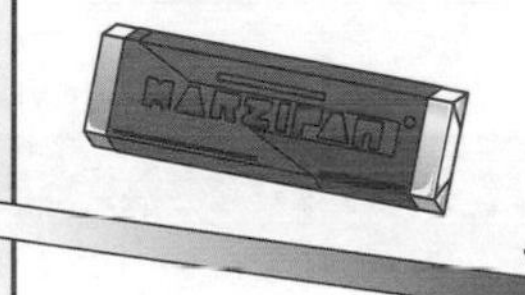

Using icing to decorate a cake is an example of a <u>finishing technique</u> (see page 58).

Standard food components often have a long shelf-life

Now you know that standard food components exist, you'll be spotting them in <u>lots</u> of different foods. Learn all these examples — you could get marks just for <u>naming</u> some standard components that might be used to make a product.

Food Preparation

The next few pages are about how to prepare meals — basically, all the techniques you need to cook.

You Should Learn How to Cook **Staple Foods**

Staple foods are foods that can be used all year round and make up the main part of people's diets.
In the UK they include bread, pastry, rice, pasta and potatoes.

BREAD is usually made by baking a dough of flour, yeast, water, salt and fat.

1) Sift flour and salt into a bowl.
2) Rub in the fat (see page 52) to make a crumbly mixture.
3) Mix yeast and warm water together and add to the rest of the ingredients. Mix to form a dough.
4) Knead the dough (see page 52) — this helps the gluten in the dough (see page 23) develop and make the dough stretchy. Cover the dough and leave to rest for at least an hour. The yeast will release carbon dioxide and make the bread rise.
5) Then you just need to shape your dough and bake it.

Pastry is made in a similar way, but it doesn't need kneading or contain any yeast.

RICE AND PASTA are cooked by boiling or steaming them so they absorb water (see pages 55-56).

1) Cook pasta until it's 'al dente', which means firm to bite but not hard.
2) Wash rice before cooking to remove starch grains on the outside of the rice.
3) Don't stir the rice as it's cooking, as this releases starch molecules and makes the rice sticky.

POTATOES can be boiled, baked, roasted, mashed or fried.

You just need to wash them and remove any green bits or 'eyes' which may contain toxins. You can peel them if you prefer, or not peel them if you like the skins (which are a good source of fibre and vitamin C). Big chunks of potato take longer to cook.

It's Good to ***Season Food*** *When Cooking*

Seasoning food means adding ingredients to improve the taste and to bring out other flavours. You can:

- rub in salt and pepper (e.g. meat) — this improves the flavour and makes the outside brown and crispy
- add a knob of butter (e.g. asparagus) after you've grilled it — this makes it look shiny and adds flavour
- glaze (see page 58) in honey (e.g. carrots) — this gives them a slightly different taste
- marinade (e.g. meat and protein alternatives) — this adds flavour by giving the food time to soak up flavours from a marinade (a liquid seasoned with herbs, spices, vinegars or oils, etc.) before cooking

Seasoning is particularly important with protein alternatives (like tofu), as they don't have much flavour.

Meat *and* ***Fish*** *are often* ***Cut*** *and* ***Filleted*** *Before Cooking*

1) You can cook some meat whole, e.g. chicken, turkey or suckling pig.
2) Cuts of meat from larger animals are usually prepared by butchers, e.g. pork chops, steaks, ribs.
3) Meat can be processed before cooking, e.g. it can be minced and made into sausages or burgers.
4) Fish can also be cooked whole, but the main bones are usually removed first.
5) You can get processed fish, e.g. fish fingers, or fillets of fish (strips of fish without the bones).

Staple foods are basic foods available all year round

All this is useful for your kitchen exploits and your exam questions. The more you practise making stuff, the more you'll remember. Plus, you'll be a better cook and impress your friends. Better get cooking.

Baking

There are several cooking techniques you should know how to do. First up, baking.

Heat can be Transferred in Three Different Ways

Transferring heat energy means moving it from one place to another — it's how your food manages to be piping hot even though you don't set fire to it (usually). Heat transfer happens in three different ways:

CONVECTION — transfer of heat energy through gases (e.g. air) or liquids.
CONDUCTION — transfer of heat energy through solids.
RADIATION — transfer of heat energy through waves of radiation.

Foods change as they're cooked — which can change a food's texture, flavour and smell. In general:

1) Solid fats in the food melt, making the inside of the food moist and tender, e.g. beef steaks.
2) Water in the food evaporates, so the food gets drier. If cooked using dry heat, proteins and sugars on the surface of the food get hot enough to go brown and crisp, e.g. bacon, roast potatoes.
3) Proteins can become firm, e.g. meat goes tough, and proteins in eggs coagulate (see page 26). If meat is cooked slowly, some proteins melt and help make it tender.
4) Starch molecules start to swell and soften, e.g. rice and pasta swell when they're cooked.
5) Nutrients (e.g. vitamin C) are broken down by high temperatures or are lost from the food.

Baking uses Dry Heat to Cook Food

1) Baking cooks food using dry heat, usually in an oven. Heat energy is transferred around an oven through radiation and convection, and through the food by conduction.
2) Because hot air rises, the top of an oven is often hotter than the bottom — that's why food cooks quicker on the top shelf than on the bottom shelf.
3) Modern ovens are usually fan-assisted or convection ovens — they have a fan inside that helps to circulate the hot air around the oven. They're much more useful because...
    - food bakes more evenly because all parts of the oven are at a similar temperature
    - the oven heats up quicker and your food cooks quicker — so they use less energy.

Using a baking tray helps to conduct heat.

Types of food that you'd usually bake include:

- bread, pastries, cakes, pies and tarts
- potatoes
- whole fishes, like sea bass or salmon
- protein alternatives, like Quorn™
- meat, e.g. in meatloaf or casseroles, although large pieces of meat are usually roasted (see page 53).

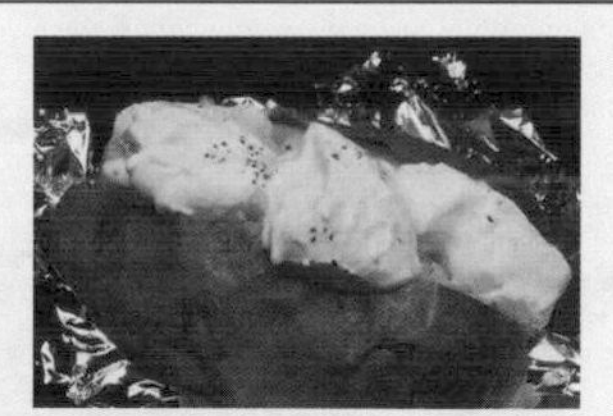

Baking food like fish or potatoes inside foil helps to keep the moisture in, making them nice and tasty.

Advantages of baking...

- It's quite healthy because no extra fat is added, and fat often leaves food as it's baked.
- The outside of the food browns and crisps, which looks and tastes nice.

Disadvantages...

- Baking food can take a long time.
- It uses a lot of energy, e.g. a large oven is kept at a high temperature for a long time.
- Food can become dried out.

Baking uses no extra fat, but it can take a long time to cook food this way

Ahhh, there's nothing quite like the smell of baking wafting through the house... Apparently, a good way to remember something is to associate it with a smell. So get cooking and revising — perfect.

Preparing Foods for Baking

There are loads of different ways you can prepare products you're going to bake.

Different **Mixing Methods** can be Used in Food Production

Most mixing methods are not just about mixing the ingredients, but also getting air into the mixture to help it rise — and give the product a nice texture.

Creaming

This is mixing fat and sugar together into a creamy mixture. It helps trap tiny air bubbles in cake mixtures, to make them rise and give a nice texture when baked.

Kneading

This means stretching and pulling dough with your hands, e.g. when making bread or pizza bases. It helps develop gluten in flour, which helps give bread its texture and allows it to rise.

Whisking

This means using a whisk or a fork to mix ingredients together and add air. E.g. whisking eggs to mix the yolk and egg white.

Folding

This means folding the mixture over, like you're folding it in half. Use a spoon or spatula to fold a cake mixture — this helps stop air being lost from the mixture while mixing.

Rubbing in

This means rubbing fat (e.g. butter) into flour using your fingertips, until the mixture looks like breadcrumbs. This adds air to the mixture and helps to make a light product, e.g. light pastry.

All-in-one

This means mixing all your ingredients together at the same time. This saves you time, but the final product may not taste as nice or rise as well as if you used separate methods.

You May Need to **Cut**, **Shape**, **Roll** or **Melt** Foods During Production

Cutting

This is using cutters (metal or plastic shapes) or a knife to make the right shapes, e.g. cut out star-shaped biscuits with a cutter.

Rolling

This is rolling out products, like pastry or biscuit dough, to the right shape and thickness. Use a rolling pin to get an even thickness, and use flour to stop the mixture sticking to surfaces.

Shaping

This is making your dough the right shape and thickness by shaping it, e.g. with your hands.

Melting

Melting chocolate or butter can make it easier to add it to a mixture. Use a microwave or put it in a bowl over a pan of hot water.

For a product to turn out well you need to use the right techniques

There are lots of things to remember on this page but most of them should be familiar from your own cooking. Cover the page and see how many methods you can remember.

Roasting and Grilling

Hopefully, you're now confident about baking, so let's get straight on to some other cooking techniques.

Fat is Added to Food When it's Roasted

1) Just like baking, roasting uses dry heat to cook food, usually in an oven. Roasting food is done at a higher temperature than baking, so foods cook more quickly and brown more.
2) Heat energy is transferred through radiation and convection, and through food by conduction.
3) Fat is often added to the outside of the food to help it to brown and stay moist. You can use extra fat or fat that's melted from the food as it's cooking. E.g. when roasting meat you can take fat that's dripped off the meat and pour it back on top — this is called basting.

Common types of food that are roasted:
- large cuts of meat, like a leg of lamb, a cut of beef or a whole chicken
- potatoes and root vegetables
- chestnuts

Advantages...

1) Extra fat and a high temperature helps to brown and crisp the outside of food, which looks and tastes good.
2) The fat from roasted meat can be used to cook other food, e.g. potatoes or fried bread.
3) Roasted food can be nice and moist.
4) Roasting can produce meat with a rare (undercooked) centre, which a lot of people like.

Disadvantages...

- Roasted food isn't always that healthy, as extra fat is often added.
- Just like baking, it takes a long time to roast food and uses a lot of energy.

Fat can Drip Off Food as it's Grilled

1) Grilling uses a dry heat source to cook food, at a higher temperature than baking or roasting.
2) Heat energy is transferred to food through radiation, e.g. heat radiation from an electric grill, or hot coals when barbecuing. Grilling using a griddle pan uses conduction to cook food.
3) Fats drip out of the food as it's grilled, and the outside of the food becomes golden and crisp.

Grilling is great...

- Food cooks quickly at a high temperature.
- It's fairly healthy as fat drips off the food.
- Just like roasting and baking, the golden outside of the food looks and tastes nice, and can have a lovely crispy texture.

But...

- It's hard to tell if the food is cooked all the way through — because the outside cooks much quicker than the inside.
- It's easier to burn food because you're cooking at high temperatures.

You can grill many foods, such as:
- smaller bits of meat like steaks, sausages
- vegetables, e.g. courgettes, aubergines
- cheeses like halloumi or goats' cheese

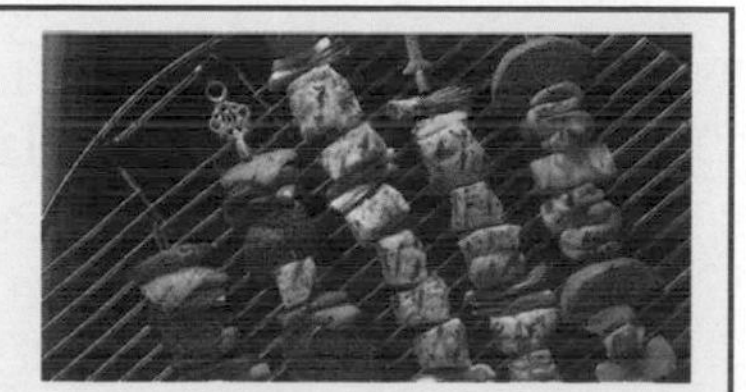

Roasting can add fat to foods but foods lose fat when grilled

Cooking using fat isn't the healthiest thing in the world. If you're trying to be healthy, grilling is a great alternative to frying (see next page) — it's well worth considering next time you do a fry-up (grill-up).

Frying

There are many different types of frying — you can choose which to use depending on the type of food you want to cook

Frying uses Hot Fat or Oil to Cook Food

Frying uses fat or oil heated to a very high temperature to cook food. Some fat is absorbed by the food as it cooks, which adds a lot of flavour. Heat is transferred to the food by convection (from the fat to the food) and conduction (through the food). Different types of frying use different amounts of fat:

The quicker food is cooked and the less fat that's used, the less fat will be absorbed — so the healthier the food will be.

Stir-frying

1) Stir-frying uses a wok coated in a small amount of oil.
2) Food cooks very quickly and needs to be moved around the wok all the time, so it doesn't burn.

Stir-fried foods include:

- Asian dishes like stir-fried noodles
- Vegetable stir-fry

Shallow frying

1) Shallow frying uses a frying pan coated in a medium amount of fat or oil.
2) Solid fats in the food melt into the pan, increasing the amount of fat in the pan.

Shallow-fried foods include:

- meat like chops, bacon and sausages
- eggs and pancakes

Deep-fat frying

1) Deep-fat frying involves completely covering food in very hot oil or fat.
2) Food is often dipped in batter (a water and flour mix) before frying — the batter absorbs the fat and goes crispy, whilst the food inside cooks nicely and keeps its moisture.

Deep-fried foods include:

- fish and chips
- doughnuts

Frying is good because...

- It's quick because the oil/fat heats up quickly.
- Food is tasty because of the flavours absorbed from the fat.
- Frying uses less energy than roasting and baking.

...but it's bad because...

- It's hard to tell if the food is cooked all the way through.
- It's easier to burn food because it's cooked at high temperatures.
- Hot oil or fat can cause serious burns, and it's highly flammable — so using a large amount can be dangerous.
- Deep-fat fried foods have more of a different type of fat in them (trans fats), which are pretty unhealthy.

Different types of frying use different amounts of oil

Although it uses fat or oil, frying isn't always unhealthy — for example, stir-frying is generally a healthy way to cook food. You can make frying healthier by using healthier fats, e.g. by using olive oil instead of butter.

Boiling

If you can boil an egg, you might be thinking that you know everything there is to know about boiling — but, just like with the other techniques in this section, boiling has good points and bad. And you need to know them all.

Boiling uses Hot Liquid to Cook Food

1) Boiling involves cooking food by heating it in a pan of boiling liquid, usually water. Simmering is like boiling, but it's more gentle, with only a few small bubbles. The liquid's at a temperature slightly lower than its boiling point.
2) Heat energy is transferred to the food through convection (from the liquid to the food) and conduction (through the food).
3) Boiling is quite a harsh method of cooking and can't be used on delicate foods, as the bubbles can break up the food.
4) Because it's boiled in liquid, the food stays moist.

If you just plonk a bit of fish into boiling water it'll probably fall apart, but you can boil fish in a bag to help protect it so this doesn't happen.

Types of food that are usually boiled include:

- tougher cuts of meat, like gammon or mutton
- potatoes and other vegetables
- rice and pasta

Advantages of boiling...

- It's quick and simple.
- Some flavours dissolve into the water, so you can use it to make a nice stock or gravy.
- No fat is added, so it's quite healthy.
- Low simmering of tough meat for a long time makes it much more tender, and so nicer to eat, e.g. braising steak in a stew.
- Boiling foods in a small amount of water with a lid covering the pan uses less energy than some other methods, e.g. steaming (see next page).

Disadvantages...

- If vegetables are boiled for too long their colour, flavour and vitamins are lost into the water.
- It's easy to over-boil food and make the texture too soft, e.g. when boiling pasta.
- Boiling often doesn't make food taste or look as nice as using other cooking methods.

Boiling is easy to do and healthy too, but it's not suitable for delicate foods

The longer you boil food, the more nutrients are lost — so get used to enjoying nice, crunchy veg. You'll use less energy that way, so everyone wins. Of course, if you eat it raw, you don't use any energy.

Steaming and Microwaving

Just two more cooking methods to go...

You Can *Steam* Food Too

1) Steaming means cooking food with steam from boiling water or stock.
2) Just like in boiling, the heat energy in steaming is transferred through conduction (from the pan to the water and through the food) and convection (from the steam to the food).
3) It's quite a gentle way to cook — so it's good for delicate food (like fish), but not for tough meats.
4) The food stays moist and doesn't become crisp.

Food you can steam includes:

- fish, like salmon or trout
- rice
- vegetables

Steaming is great...

- It's dead simple — if you have a steamer.
- No fat is added, so it's a healthy way to cook.
- Lots of food can be steamed at the same time in stacked pans.
- Veg keep more of their taste, texture, colour and nutrients.
- You don't drain food so it's less likely to break up.

However...

- Food may not have as much flavour.
- It takes a lot of energy to produce enough steam to cook the food.

Microwaves Use *Radiation* to Cook Food

1) Microwaves heat up the water, fat and sugars in the food to cook it.
2) Heat energy is transferred to the food by microwave radiation (surprisingly enough).
3) Convection microwaves combine the advantages of a microwave with those of a convection oven — so you can cook with the speed of a microwave, but food still comes out crispy and brown like it was baked in an oven.

Advantages of microwaving food...

- You can defrost, reheat and cook food quickly.
- Food is healthier — more vitamins remain in the food.
- Microwaving is efficient — it uses less energy than other cooking methods, especially for small portions.

Disadvantages...

- Eggs and closed containers can explode.
- Not all cooking equipment can be used, e.g. you can't put metal in a microwave.

You can microwave foods like:

- fish and chicken
- puddings, e.g. sticky toffee pudding
- soups
- ready meals

Steaming and microwaving are healthy ways to cook...

...but unfortunately, using a healthy method of cooking doesn't automatically make food healthy, e.g. microwaving a salt and fat-filled ready meal doesn't make it healthy, no matter how much you may want it to.

Sauces and Soups

There are loads of different techniques you can use to make your own perfect soup or sauce.

Sauces can be Made in Lots of Different Ways

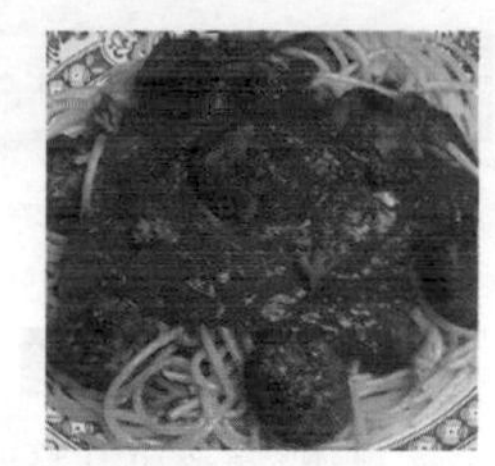

Sauces are used in both savoury and sweet dishes — as an accompaniment (e.g. custard, ketchup) or as part of the dish (e.g. in a lasagne). You can put pretty much anything you like in a sauce, but they usually contain at least one of these: flour, butter, eggs, cornflour, stock, milk, cream. There are three main ways to make a sauce:

1. You can make a **roux** as a base for your sauce, by melting some butter in a pan and mixing in an equal amount of plain flour. Stir until the flour's cooked. You can use the roux to make a sauce by gradually adding a liquid, e.g. milk, stirring to avoid lumps the whole time. Simmer, then add the rest of your ingredients. The roux will help to give your sauce a good consistency (thickness) and flavour.
2. You can **blend** your ingredients using a food processor. By blending them for different lengths of time, you can get the consistency and texture you want — from a chunky guacamole to a smooth ketchup.
3. Or you can just make your sauce **all-in-one** by putting all your ingredients into a saucepan. Whisk over a medium heat, until your sauce starts to bubble and thicken, then leave it to simmer for about 5 minutes.

Soups can Contain Pretty Much Any Type of Food

Soups are great — you can add almost any ingredient you like and they're really easy to make. Like sauces, there are some key ingredients that most soups contain:

1) Stock — a liquid base to start the soup.
2) Plain flour, cornflour or a roux — to thicken the soup.
3) Potatoes or onions — to thicken and provide bulk.
4) Salt, pepper and spices — to season and flavour the soup.
5) Cream or butter — to make it taste creamy and delicious.

There are different methods of making soups:

The all-in-one method, e.g. to make leek and potato soup
Cook all your ingredients together in some stock, over a low heat. Use a blender to get the right consistency, e.g. smooth or lumpy, then add seasoning and cream to taste.

Adding your ingredients one at a time to make a broth, e.g. scotch broth
Boil some meat on the bone to produce a stock, then boil some vegetables in this. Scrape the meat off the bones when it's cooked and add this in to make your broth — a liquid with chunks of food in it.

Desserts can be Uncooked, Baked or Chilled

You should really sample different types of desserts to revise this. But in the mean time, check this out:

1) Baked desserts — you cook these in an oven, e.g. cakes, pastries, biscuits, scones, apple crumble.
2) Uncooked desserts — these are things like fruit salad or jelly, which don't need to be cooked.
3) Some baked or uncooked desserts are served chilled, e.g. cheesecakes, yoghurt or ice-cream.

A roux is useful as a base for a sauce or soup

There's loads to learn here but take your time and you'll get there — and it could have tasty results both in the kitchen and in your exam. Try out the different cooking methods — it'll help you remember them.

Finishes and Cooking Efficiently

Finish off the first part of this section with some finishes.

Finishing Techniques Make Food Look *More Appealing*

You can use different techniques to put the finishing touches to your food and make it look nice — food often seems to taste better when it's nicely presented.

1 Garnishing

This is when you add extra little bits of food before serving your dish to add extra colour or flavour. E.g. chocolate flakes or hundreds and thousands on ice-cream, a drizzle of cream and some chopped herbs on a bowl of soup, or some grated parmesan and black pepper on top of pasta.

2 Glazing

This is when you apply a shiny coating to the top of food. This is often sweet, e.g. sugar and water glaze on doughnuts.

3 Decorating

This covers tons of stuff, like fancy icing on wedding cakes, putting a swirly pattern of sauce around the outside of your plate, or even making your cakes into swan shapes.

You can *Reduce* the *Energy* You Waste When You Cook

Energy is always wasted when you cook food. But simple steps can help reduce the amount of energy that's wasted, which is good for the environment (and for your parents' pockets).

1) Only boil the amount of water you need — boiling water in a pan or a kettle uses an awful lot of energy. So only boil what you need, e.g. just enough water to cover your rice.

2) Use an appropriate size of pan and ring — if you're only boiling one egg, don't use the biggest saucepan on the largest ring on the hob because this wastes energy (and it'll take longer to cook).

3) Cover saucepans and pots with lids — this stops heat energy being lost to the kitchen.

4) Cook food together — cook vegetables all together in the same pot or steamer. Or if there's room in the oven for the potatoes as well as the meat, put them both in together.

5) Use the most energy efficient cooking method — microwaves are very efficient (see page 56).

Finishes may not always add flavour, but they make food look good

Presenting your food in a nice way makes it look more attractive, so people are more likely to want to eat it. Makes sense, so learn the three finishing techniques above. Learn the five ways to reduce energy wastage when cooking too.

Warm-Up and Worked Exam Questions

You must be getting used to the routine by now — the warm-up questions run over the basic facts and the worked examples show you how you should answer exam questions. Then, the rest is up to you.

Warm-up Questions

1) Name one example of:
   a) a natural emulsifier.
   b) a gel.
2) a) Why do people season food before cooking it?
   b) Give an example of how you can season food.
3) Describe a situation when you might use folding in cooking, and describe what you'd do.
4) Which type of frying is the healthiest and why?
5) Why is it not a good idea to boil fish? How could you get around this problem?
6) Explain how you could use finishing techniques to decorate a birthday cake.

Worked Exam Questions

Another lot of worked questions with hints to explain how to do them.
You'll get the most out of them if you cover the answers and try them for yourself first.

1 Standard components are often used in the production of ready made lasagnes.

(a) Explain what is meant by the term 'standard component'.

A ready made ingredient.

(1 mark)

(b) Name three standard components a manufacturer could use to make a lasagne product.

Think about the ingredients in a lasagne and work out what can be bought ready made.

1. lasagne sheets
2. pre-made tomato sauce
3. pre-made cheese sauce

(3 marks)

2 One method of cooking food products is to bake them in an oven.

(a) Give two advantages of using an oven that's fan-assisted over one that is not.

The food bakes more evenly. Also, the oven heats up quicker and your food cooks quicker.

(2 marks)

(b) List three types of food that can be baked.

This is the kind of question where you can get easy marks — there are loads of foods you could put down and there's no need to go into any detail.

1. bread
2. potatoes
3. meat

(3 marks)

(c) Give one advantage and one disadvantage of baking food.

It's quite healthy because you don't usually add any extra fat, and fat often leaves food as it's baked, but it can take a long time to bake food.

(2 marks)

Exam Questions

1 A company that specialises in desserts has recently decided not to use standard components, but to use fresh ingredients instead.

(a) Give one reason why a consumer might prefer to buy a dessert that contains no standard components.

....

(1 mark)

(b) Give three disadvantages to manufacturers of using standard components.

1. ....
2. ....
3. ....

(3 marks)

2 Many types of food can be cooked by grilling.

(a) (i) How is heat transferred to food when grilling using a griddle pan?

....

(1 mark)

(ii) What type of heat transfer is involved in barbecuing?

....

(1 mark)

(b) Give one advantage and one disadvantage of grilling.

Advantage ....

....

Disadvantage ....

....

(2 marks)

3 Describe three ways you can minimise the amount of energy used in cooking Spaghetti Bolognaise.

....

....

....

....

....

....

....

(6 marks)

Scale of Production

Manufacturers use different types of production depending on how many of a product they're making.

One-Off Production — Every Product's Unique

1) This is where you make a single product (it's also called 'job production').
2) Every product's made differently to meet a specific request, e.g. a wedding cake.
3) Every product needs an individual recipe and an individual method.
4) Experienced workers with specialist skills are needed.
5) The products are high quality but they take a lot of time and cost a lot.

A prototype — something you get people to sample to see if they would buy it, before developing it further — is a one-off product.

Batch Production — Specified Quantities of a Product

1) This is where you make lots of your product in one go — each load you make is called a batch.
2) Every batch is made to meet specific requests from retailers, e.g. 100 chicken pies.
3) You can change batches to make another similar product, e.g. 500 steak pies.
4) But machines need to be cleaned between batches — this 'down time' is unproductive. Staff and machines need to be flexible so batches can be changed quickly, to avoid losing too much money.
5) Batch production is quicker than one-off production and it's a bit cheaper.

Mass Production — Large Quantities of a Product

1) This is where you make large numbers of a product that sells well, e.g. a loaf of sliced bread.
2) The product's made using a production line — it passes through various stages of production. Products are made very quickly so they're cheap.
3) Machines are used at some of the stages so fewer workers are needed.
4) But to make a new product you need to change the production line — this can take a long time and this unproductive time costs money.

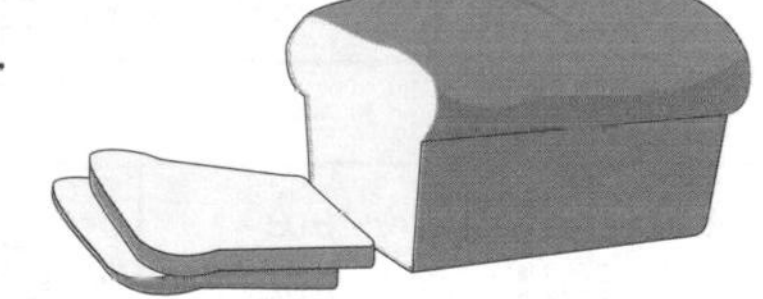

Continuous Flow — Non-Stop Production 24hrs/day

1) This is where you make a product all the time, with no interruptions. It's basically non-stop mass production, with a production line using expensive, specialised equipment.
2) It's used for products that are sold regularly and in large numbers, e.g. baked beans.
3) It'd cost too much to keep turning equipment off and then re-starting it — so everything runs all the time. This keeps production costs really low.
4) But if anything goes wrong it takes time to get it going again, and unproductive time costs money.

Learn the four types of production

As usual, it's all about money. It's much cheaper for industries to produce something in huge numbers than one at a time — but it only works if the products needed are all the same and if they sell really well.

Scale of Production

Computers are extremely useful for designing and manufacturing food products.
They can make the production process easier and quicker — and much safer.

CAD Helps to Design Products...

1) This is where computers are used to help design a product.
2) You can use computer-aided design (CAD) to produce models of the product and its packaging in 2D and 3D — so you can view them from any angle.
3) Once the product's been drawn on screen, you can easily recalculate values and change the design until you're happy. CAD is more accurate and much quicker than drawing and re-drawing your designs on paper.
4) CAD is really useful to calculate things like your product's nutritional value (see page 8) portion size, shelf-life and what it'll cost to make.

...and CAM Helps to Manufacture Them

1) Computers are used in the manufacturing process too — it's called computer-aided manufacture (CAM).
2) Computers control some or all of the production stages, e.g. they're used to weigh out the correct amount of each ingredient, set the correct oven temperature and cooking time, etc.
3) The whole production process is overseen by someone who keeps a close eye on everything.

CAM has lots of advantages:

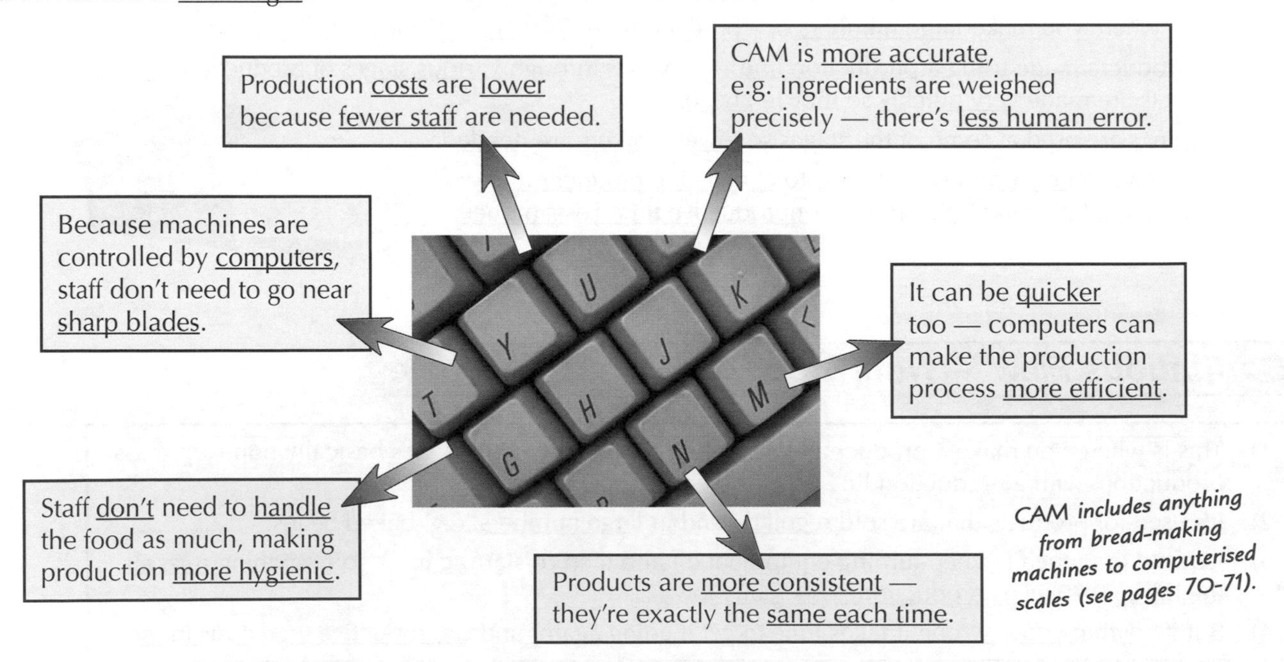

CAM includes anything from bread-making machines to computerised scales (see pages 70-71).

CAD and CAM make production more efficient

This page is basically about how and why CAD and CAM have aided design and manufacture in industry. They've had a massive effect on the speed and accuracy of production — important stuff.

Quality Control

If a product is high quality, customers are more likely to buy it again.

Mass-Produced Products need to be Consistent

Manufacturers who make products on a large scale aim to produce consistent products — products that are the same every time, e.g. they have the same taste, colour, portion size, etc. Customers can rely on the product being just like the one they liked before.

Using standard methods and equipment helps to produce consistent products, e.g:
1) All ingredients are weighed using accurate electronic scales so they're a consistent weight.
2) Using standard moulds, templates and cutting devices produces a consistent size and shape.
3) To get the flavour and texture consistent, standard food components can be used (see page 48).
4) Using identical ingredients, cooking times and temperatures gives a consistent colour.

Products are Checked by Quality Control

Manufacturers set standards that their product must meet and they check to make sure these standards are being met — this is called quality control. Checks are made at every stage of production and the final product is checked too.

1) Visual checks
   - The colour of the product is checked against a standard colour.
   - The packaging is checked to make sure it's not damaged and all the labels are clearly printed.
2) Testing
   - The taste is tested at the end, to make sure it's exactly what the manufacturer wants.
   - The size, thickness and pH of the product may also be tested.

Any Problems are Fed Back Straight Away

There's no point in a manufacturer doing all this checking unless any problems are then corrected.
1) If a product's not right, the problem is relayed back to the factory floor — this is called feedback.
2) Feedback happens straight away so the problem can be fixed quickly.
3) This means ingredients aren't wasted — so the manufacturer doesn't waste time and money.

EXAMPLE — biscuit production

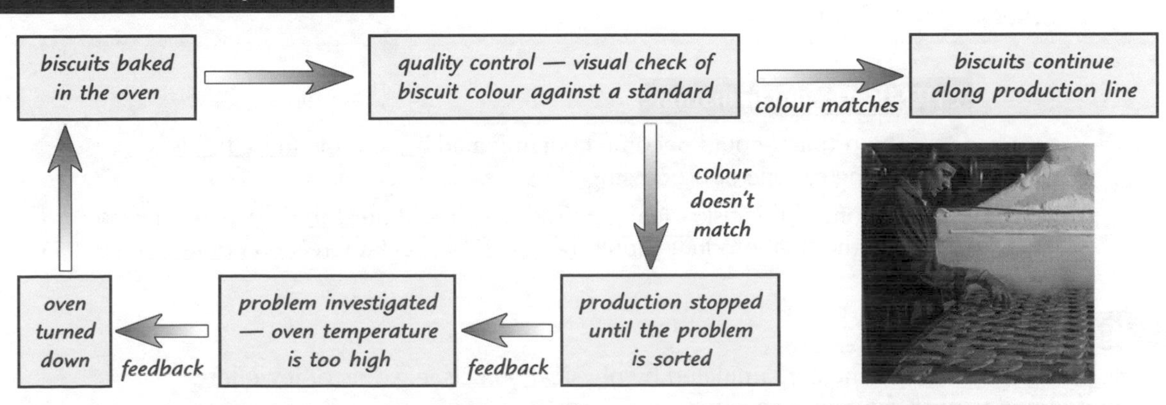

Good quality control means you'll produce consistent products

It makes sense to check that a product's right before you spend tons of money making loads more. Quality control is about checking you're meeting standards and doing something about it if you're not.

HACCP

Risk assessment and HACCP are important in food production — and they might just come up in your exam.

Risk Assessment is all about Hazards and Risks

HAZARD

Anything that could cause harm or problems during the making, packaging, storing or transport of a product is a hazard.

RISK

The risk is the likelihood of that hazard actually causing a problem.

RISK ASSESSMENT

Risk assessment means thinking about: what could happen, when it could happen, and what steps are needed to reduce the risk. It applies to both food hygiene and the safety of workers.

HACCP Helps Avoid Food Contamination

It's important that products won't harm the consumer. HACCP helps identify potential problems and put controls in place to prevent food being contaminated before it reaches the consumer.

Hazard Analysis Critical Control Points (HACCP)

- A HAZARD is anything that's likely to cause harm.
- ANALYSIS is when you look in detail at something.
- CRITICAL means it's very serious.
- A CONTROL POINT is a step in the process where you put in a control to prevent a problem from occurring.

HACCP tries to stop Three Types of Contamination

BIOLOGICAL contamination

The product could become contaminated by bacteria, especially if it contains high-risk foods (see page 66). E.g. there's a risk that eggs could carry salmonella. To control this risk, random samples of eggs would be tested for salmonella near the beginning of the production process. You could also take samples of the end product, e.g. quiche, to be extra cautious.

CHEMICAL contamination

1) The product could become contaminated by, say, cleaning fluids during storage or processing.
2) To control this risk, cleaning products should be stored away from food and the final product should be tested to check there's no contamination.

PHYSICAL contamination

1) The product could become contaminated by physical objects, e.g. bits of jewellery, chipped nail varnish, hair, insects, etc.
2) To control these risks, workers wear overalls and hairnets, with no jewellery or nail varnish allowed, and food is kept covered as much as possible. Finished products are also checked.

HACCP

Set up HACCP Step by Step

1) In rough, write down how you're going to make your product in a series of steps.
2) Think about your product from 'field to table', i.e. consider the stages your ingredients will go through, from the beginning (before they're harvested) to the end (when the product's bought by the consumer).
3) Use a simple flow chart to set out the steps that you've identified.
4) Then consider any potential hazards associated with each step. For example:

   a) The production or purchase of the ingredients — you need to make sure you get high quality ingredients from a reliable source.
   b) Storage of ingredients — e.g. make sure dry ingredients aren't kept in damp conditions.
   c) The actual making process (this is the most important area for HACCP).
   d) Packaging — e.g. make sure the packaging doesn't damage the product.
   e) Transport from factory to shop — check it's being transported at the right temperature and won't get damaged.

5) Finally, you have to think about how you can control and prevent problems from taking place.

Example: HACCP for a Decorated Cake

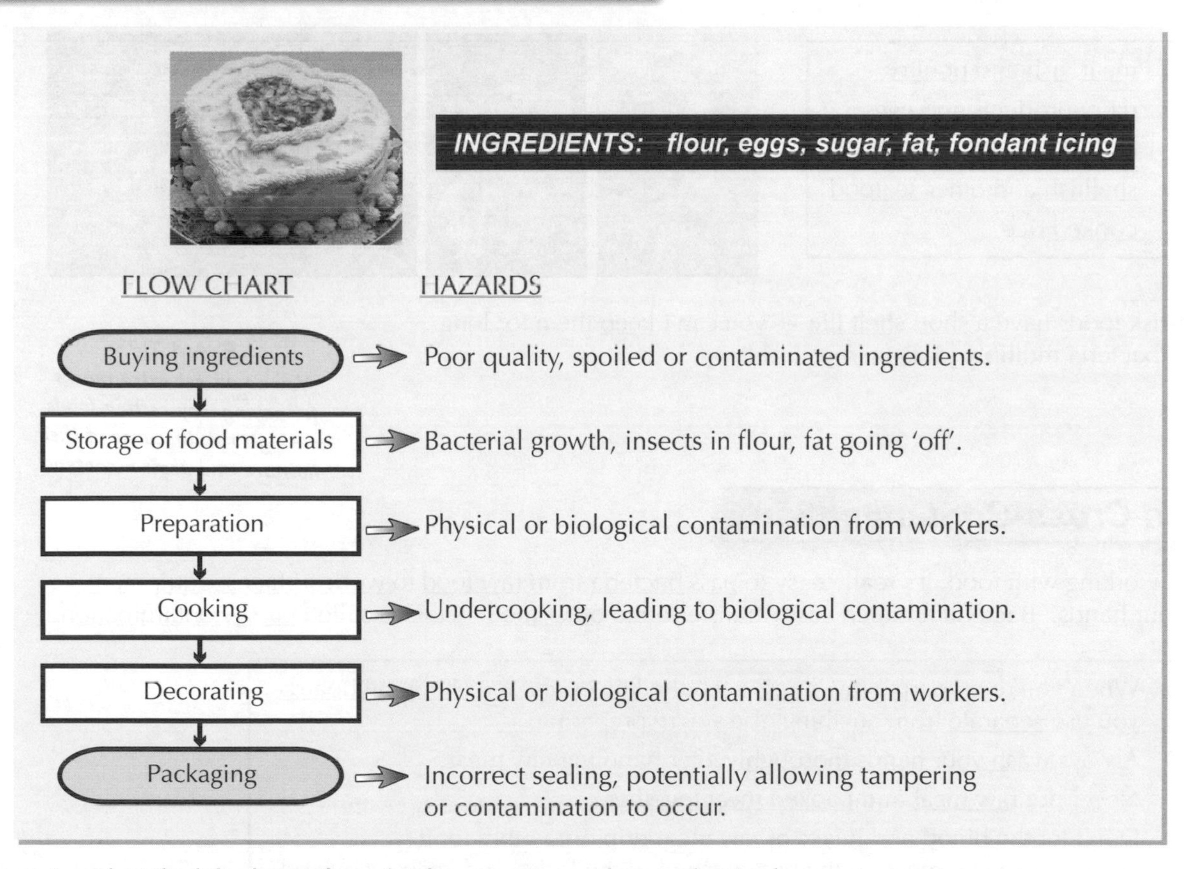

Once you've identified the hazards and risks, you can work out what to do — how to stop them or control them. These would be your control points.

Food can be contaminated in three different ways

There's a whole pile of words and initials to remember here, but the principle is pretty simple. Identify hazards, work out the risk of these hazards happening, then try to reduce these risks.

Food Contamination and Bacteria

If people eat food that's contaminated by biological, chemical or physical hazards they could become very ill. So you have to handle food safely and hygienically.

Bacteria are the Main Cause of **Food Poisoning**

1) The symptoms of food poisoning include sickness, diarrhoea, stomach cramps and fever. In extreme cases, especially where people are old or vulnerable, it can lead to death.
2) The main cause of food poisoning is eating food (or drinking water) that's contaminated by bacteria. Bacteria are found in air, water, soil, people, animals — pretty much everywhere, really.
3) You can't see bacteria — they're so small you've got to use a microscope to spot them.
4) They often don't make the food look, taste or smell any different — so it's hard to know they're there.
5) Bacteria like conditions where they can multiply very quickly. These include:

- *moisture*
- *warmth*
- *neutral pH*

Food poisoning can also be caused by chemical or physical contamination — see page 64.

Bacteria Grow Fastest in **High-Risk** Foods

High-risk foods are foods that bacteria grow quickly in, because they're moist and high in protein. High-risk foods include:

1) meat, fish and poultry
2) dairy products and eggs
3) gravies, stocks and sauces
4) shellfish and other seafood
5) cooked rice

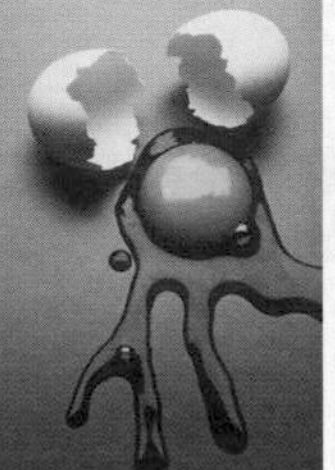

High-risk foods have a short shelf life — you can't keep them for long, or the bacteria multiply to dangerous levels.

EXAM TIP
You could get extra marks for saying why certain foods are high risk (because of their moisture and protein content).

Avoid **Cross-Contamination**

When working with food, it's really easy to pass bacteria from raw food to work surfaces, equipment and your hands. Bacteria are then easily transferred to other food — this is called cross-contamination.

1) When you're preparing raw meat, keep the knives and the chopping boards you use separate from anything else you're preparing.
2) Always wash your hands thoroughly after handling raw meat.
3) Never put raw meat and cooked meat together.
4) Don't let the blood and juices of raw meat drip onto other food — always store raw meat on the bottom shelf in the fridge and keep it covered.

High risk foods are high in protein

Knowing all this stuff will help you through your GCSE and it'll save your stomach loads of grief — so it must be worth it. Learn the examples of high-risk foods and how to use them safely.

Food Contamination and Bacteria

The chance of food contamination can be reduced by simple procedures.

Follow Safety and Hygiene Procedures at Every Step

PURCHASING FOOD

- Always buy food from a reputable supplier so you know it's good quality.
- Take note of the use by date and make sure you can use it before this date.
- Check food carefully, e.g. make sure it hasn't been squashed or gone mouldy, check the packaging isn't damaged and the seal is still intact.

STORING FOOD

- Always follow the storage instructions — especially about temperature.
- Use old purchases before they go out of date.
- Keep food sealed or covered up.

PREPARING FOOD

- Follow personal hygiene procedures — wash your hands, wear a clean apron, wear a hat or hair net, remove all jewellery, cover all cuts, report to the person in charge if you're ill, don't taste food with your fingers.
- Always use clean equipment.
- Avoid cross-contamination (see previous page).
- If you're defrosting frozen food before cooking it, make sure it's defrosted fully.

COOKING FOOD

- Cook food at the right temperature (see below) and for long enough.
- Make sure food is cooked all the way through, e.g. cook thicker pieces of meat for longer than thin ones — test the middle to make sure it's cooked properly.

SERVING FOOD

- Serve hot food straight away.
- If you're serving food cold or storing it, cool it down as quickly as possible. Keep food covered so it's away from flies — preferably put it in the fridge.

Cook and Reheat Food to the Right Temperatures

1) Cook food thoroughly to kill bacteria — the temperature should be 72 °C or more in the middle.
2) If you're keeping food warm, keep it at about 70 °C, and don't keep things warm for more than an hour before eating.
3) If you're reheating, make sure the food is heated to at least 72 °C for at least three minutes.

An EHO Checks Things are Safe and Hygienic

1) Environmental Health Officers (EHOs) maintain and improve public health standards — things like checking food hygiene and making sure health and safety regulations are being followed.
2) In the food industry, this means checking that food storage, preparation and sale areas are clean and safe to work in, and that the food produced is safe to eat.

Keep your kitchen clean to help avoid food contamination

As you probably know, food poisoning is pretty horrible — and that's why EHOs are so important. They make sure that the food industry sticks to good safety and hygiene procedures.

Preservation

Cooking and storing foods at the correct temperatures extends their shelf life and keeps them safe to eat for longer.

The Right Temperature is Vital When Preserving Food

To preserve food, you need to keep it in conditions that bacteria can't grow in. First up, there are some critical temperatures that affect bacterial growth:

Cook food ABOVE 72 °C to KILL bacteria

These preservation methods all use heat:

1) Canning — food is put into a sealed can and heated to 115 °C, killing any bacteria. Because the can is sealed, no more bacteria can get in.
2) Bottling — this is like canning but with bottles.
3) Pasteurisation — food, e.g. milk, is heated to 72 °C then carefully packaged to make sure that no bacteria contaminate the food.

COOK CHILL products last for up to 5 days

1) Products are cooked then chilled (to between 0 °C and 3 °C) within 90 minutes.
2) They're stored in a fridge (between 0 °C and 5 °C) for up to 5 days.
3) They should be reheated (to above 72 °C) before being eaten. They can't then be reheated again.

°C 80 70 60 50 40 30 20 10 0 –10 –20

The DANGER ZONE is 5 to 63 °C

1) Bacteria grow and multiply quickly in temperatures from 5 °C to 63 °C — this range of temperatures is called the danger zone.
2) The optimum temperature for bacteria growth is 37 °C.

Freeze food at –18 °C to STOP THE GROWTH of bacteria...

1) Freezing food at –18 °C or lower stops bacteria growing — they become dormant.
2) Freezing greatly extends the shelf life of the food and the nutrients aren't lost.
3) It doesn't kill the bacteria though. They become active again when the food defrosts.

Chill at 0 °C to 5 °C to SLOW the growth of bacteria

1) Keeping food in the fridge, between 0 °C and 5 °C, slows down the growth of bacteria.
2) This extends the shelf life of the food — although it won't last as long as canned or bottled foods do.
3) High-risk foods MUST be kept chilled to prevent the risk of food poisoning.
4) Chilling food doesn't change its properties much — chilled food looks and tastes the same — but it may have a harder texture.

...and to preserve COLOUR and FLAVOUR

1) Vegetables contain ripening enzymes that make them go brown when they're stored for long. To stop this you can blanch them — plunge them into boiling water to kill the ripening enzymes — and then freeze them. That way they'll keep their colour.
2) Accelerated freeze drying means quickly freezing food and then drying it in a vacuum so that the ice turns to water vapour. This method is used for instant coffee and packet soups because it keeps the colour and flavour.

Bacteria thrive between 5 and 63 °C — the danger zone

This stuff is really important to keep food safe to eat. Cover the page and check whether you can remember what the critical temperatures are for each preservation process.

Preservation

Additives also help to preserve food, as well as enhancing colour and flavour.

Don't Let Food go Past its Best

Use by date

1) The use by date is shown on products with a short shelf life, e.g. high-risk foods (see page 66).
2) It's given as a safety warning. If you use the food after this date, it might not be safe — you run the risk of getting food poisoning.

Best before date

1) The best before date is shown on products with a longer shelf life, e.g. tinned foods.
2) It's given as a warning about quality. If you eat the food after this date, it's probably safe but might not be as nice as you'd expect, e.g. biscuits could be soft.

Chemicals can Preserve Food but they Alter the Taste

Bacteria can't grow if you use certain chemicals to preserve food — but this changes the taste

- Salt — salt absorbs water from bacteria, making them shrivel up and die. Salt is used to preserve meats like ham and bacon — but it makes food taste salty.

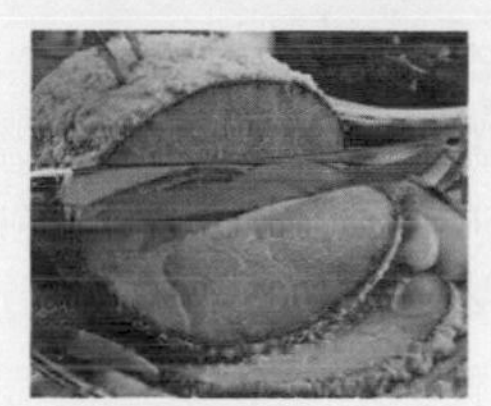

- Sugar — using high amounts of sugar (e.g. in jam) kills bacteria in much the same way. But then of course the food tastes very sweet.

Bacteria also need a neutral pH (6.6 to 7.5) to grow — so making food acidic or alkaline can preserve it:

- Vinegar — vinegar is too acidic for bacteria to grow — but it also gives food an acidic, tangy taste and can make it look brown. Vinegar is used in chutneys and pickles.
- Concentrated lemon juice — lemon juice is also acidic. It's used to preserve fruit salads.

See pages 37-38 for more on acids and alkalis.

There are Other Methods of Preservation

1) Drying — this removes all the moisture so bacteria can't grow.
2) Irradiation — food is zapped with radiation, which kills the bacteria.
3) MAP (modified atmosphere packaging) and vacuum packaging are other methods — see page 84.

Salt, sugar, vinegar and lemon juice are all natural preservatives...

...but they change the taste of the food. This can be a good thing — jam wouldn't be jam if it wasn't sweet. However, preserving fruit by putting loads of vinegar on it could taste very unpleasant indeed.

Domestic and Industrial Equipment

There are loads of different tools to make use of in food technology...

Different Tasks Need Different Tools

To make a good product, you need to be able to select the right equipment and be able to use it safely.

WEIGHING AND MEASURING

1) You need to be able to weigh and measure ingredients accurately.
2) Weighing scales (electronic and balance scales) are used to weigh dry ingredients (e.g. flour) and solid ingredients like butter.
3) Liquids are measured in measuring jugs or cups (some dry ingredients are measured in cups too).
4) You can use measuring spoons for small amounts of ingredients.

CUTTING

1) You use different knives depending on what you're cutting, e.g. bread or cheese knives.

2) A mandolin can be used to slice and cut foods evenly.
3) You can use pastry cutters to cut out the same size and shape every time.

MIXING

1) The tools you use for mixing depend on what you're mixing and the quantities you're using.
2) There are different ways of mixing ingredients — e.g. blending, folding, whisking, etc. (See page 52.)
3) For mixing fairly small amounts of ingredients, you can mix by hand using a spoon or a whisk. But mixing by hand can be hard work and take ages, and won't always give a smooth, consistent mixture.
4) Hand-held mixers and blenders save time and effort, and produce more consistent results.
5) Food processors can be used for mixing, slicing, chopping or dicing food. Using the same settings will give you the same results each time.
6) A breadmaker (or a hook attachment for a food processor) will mix and knead bread dough. This saves the effort of kneading by hand — it's also more hygienic.

SHAPING

1) You often need to be able to shape your ingredients.
2) You can do this by hand, e.g. shaping dough into biscuits. Your results won't be very consistent — there'll be lots of different shapes and sizes.
3) Cutters can help you make more consistent shapes.
4) You can use moulds to set liquids into different shapes (e.g. jelly).

HEATING AND COOLING

1) There are lots of different ways you can heat things up. You can use ovens, hobs, microwaves and steamers — it all depends on what you want to cook.
2) There are different types of thermometer for measuring the temperatures of rooms, ovens, fridges and freezers. Temperature probes are used to measure the temperature inside food, by sticking the probe right into the middle. You should use a probe with high-risk foods (see page 66) to check they're properly cooked.
3) You use refrigerators to cool things down and freezers to freeze things (fairly obvious really).

Domestic and Industrial Equipment

Electrical Equipment has lots of *Advantages*

1) Electrical equipment is any piece of equipment that works using electricity (from the mains or batteries).
2) It works the same way every time — so you get consistent results.
3) You get a quality product because of accurate measurements and precise timings.
4) It's much quicker and easier to use electrical equipment (e.g. for whisking) than doing things by hand.

EXAMPLE — electronic scales are much better than traditional scales...

1) They'll weigh ingredients accurately to within 0.05 g.
2) There's less room for human error in reading a digital display than with judging when balance pans are level (or when a needle is pointing to the right number).
3) You can preset the scales to weight different ingredients, so they're more accurate. This saves time — products which are underweight or overweight can then be rejected.

4) They can be linked to a computer so that feedback is immediate (e.g. no more ingredients are added when a certain weight is reached) — they're often used in CAM (see page 62).

There's Loads of Other Industrial Equipment too...

1) Industrial ovens are usually computer-controlled and bigger than the ones you'd use at home or at school. Examples include tunnel ovens, deck ovens and travel ovens.
2) Some food is cooked in vats — huge containers usually made from stainless steel.
3) A hopper is a huge holding container that can feed in the correct amount of ingredients.
4) A centrifuge works like a huge spin dryer — it separates liquids from solid parts, e.g. to make olive oil.
5) Depositors are huge tubes, nozzles or funnels which fill containers like pastry cases and moulds.
6) Floor-standing mixers are large food processors — they can mix huge quantities of ingredients.

Use Equipment Safely and Hygienically

1) Always read the instructions carefully before using equipment.
2) Everyone should wash their hands both before and after using equipment.
3) All equipment needs to be thoroughly cleaned after use to prevent cross-contamination (see page 66).
4) Workers should be given appropriate health and safety training — e.g. using safety guards, emergency stop buttons and regularly servicing the equipment.

Make sure you know the best tools for making your product

Manufacturers make tons and tons of consistent products, so it's no wonder they use all sorts of posh, high-tech gadgets. Learn the names of all the different pieces of equipment and the uses of each one.

Warm-Up and Worked Exam Questions

All this thinking about food may be fun, but now it's time to check out some warm-up questions to find out how much of it you've taken in.

Warm-up Questions

1) A cafe sells a range of sandwiches and usually sells about 1000 sandwiches per day.
   a) Suggest what method of sandwich production would most suit the cafe and explain why.
   b) A customer wants to order a type of sandwich that isn't on the cafe's menu, so the owner agrees to make it for him. What is this method of production called?
2) A worker checks that the flapjacks passing him on a conveyer belt are the correct size. He notices that some are too small. Explain what should happen in the feedback process to fix this problem.
3) A chef is preparing a Chicken Caesar Salad.
   a) Name a high-risk ingredient he is using and explain why it's high risk.
   b) Name three things the chef should do to avoid cross-contamination when preparing the salad.

Worked Exam Questions

These worked exam questions aren't here to aid your laziness. They're here to give you useful hints on decoding and answering exam questions. Don't rush through them — use them wisely.

1 A biscuit manufacturer sells a range of plain and chocolate-coated biscuits to local shops. The biscuits are produced using batch production.

(a) What is meant by the term 'batch production'?

Making specific quantities of a product in one go.

(1 mark)

(b) Explain why batch production is the most suitable production method for the biscuit manufacturer.

There are three marks available here so make sure you give at least three reasons.

The manufacturer only sells the biscuits to local shops so probably doesn't sell enough to make mass production or continuous flow production cost-effective. The manufacturer makes a range of biscuits so needs to change batches every so often, which can be done easily with batch production. One-off production isn't suitable as the manufacturer wouldn't be able to make one biscuit at a time.

(3 marks)

(c) The manufacturer is considering using mass production to make the most popular line of chocolate biscuits. Give one benefit of using mass production instead of batch production.

It's more cost-effective in the long-term.

(1 mark)

2 Quality control checks are vital in the manufacturing process.

(a) Explain what is meant by the term 'quality control'.

Checking that standards are being met, in order to produce a high quality product.

(1 mark)

(b) Explain why control checks are important in food production.

The manufacturer can produce consistent products. They make sure the product meets the design criteria.

(2 marks)

Exam Questions

1 Food poisoning can be caused by eating food contaminated with bacteria.

(a) Give three typical symptoms of food poisoning.

1. ....

2. ....

3. ....

(3 marks)

(b) Explain why heating food to over 72 °C helps reduce the risk of food poisoning.

....

(1 mark)

2 Chilling and freezing helps to preserve food.

(a) (i) State the two temperatures between which chilled food should be stored.

....

(1 mark)

(ii) Explain why chilling high-risk foods increases their shelf life.

....

....

(1 mark)

(b) Evaluate the advantages and disadvantages of freezing a product to preserve it.

....

....

....

....

....

(4 marks)

3 Explain why most industrial manufacturers choose to use computerised weighing equipment to make their products.

....

....

....

....

....

(4 marks)

Revision Summary for Section Three

Well done — you've finally reached the end of the longest section in the book. Now all you need to do is test how much you've learnt with these lovely revision questions.

1) What is an emulsion?
2) Explain the difference between a solution and a suspension.
3) Give three ways the nutritional value of a cake could be changed.
4) Give two advantages of using standard components.
5) What standard components might be used to make a pie?
6) What is a staple food? Give three examples of staple foods in the UK.
7) Briefly describe how to make bread.
8) How is heat transferred to the food during baking?
9) How does protein in food change during cooking?
10) What is kneading?
11) Describe one way of melting butter.
12) a) How is roasting different from baking?
    b) Name a food that can be roasted.
13) Give two examples of foods that can be grilled.
14) a) Name the three types of frying and give an example of a food commonly cooked by each type.
    b) Give an advantage of frying food over roasting it.
15) a) How is boiling different from simmering?
    b) Describe how heat energy is transferred to food during boiling.
    c) Give an advantage and a disadvantage of boiling food.
16) a) How is steaming different from boiling?
    b) Why is steaming considered to be a healthy way of cooking?
17) Give one advantage of microwaving over steaming.
18) a) What is a roux?
    b) What's the main difference between a soup and a broth?
19) Give an example of a baked dessert.
20) a) Describe how you could garnish a nice pasta meal. Why might you want to do this?
    b) What is a glaze and what's it used for?
21) What is mass production?
22) a) What is continuous flow?
    b) Why might the manufacturer choose this method of production?
23) What does CAM stand for?
24) Describe how using CAD helps a manufacturer to design their product.
25) Describe:
    a) a hazard b) a risk c) a risk assessment
26) What does HACCP stand for? Why do manufacturers use HACCP?
27) Name the three different types of contamination.
28) Describe how you'd set up a HACCP.
29) Describe the three conditions that allow bacteria to multiply very quickly.
30) How can you slow the growth of bacteria? How can you stop their growth?
31) What does an Environmental Health Officer do?
32) What is meant by the danger zone?
33) How is a use by date different from a best before date?
34) a) Explain why salt can be used to preserve foods.
    b) Describe a way food can be preserved using heat.
35) Explain why it can be better to use electrical equipment rather than make products by hand.
36) a) Name a piece of electrical equipment you could use to knead dough.
    b) What equipment would you use to measure the temperature in the middle of food?
37) Name two pieces of equipment that industrial food producers use, and say what each is for.

Social Issues

Manufacturers don't just make a product and hope people will buy it. They look for groups of people with specific needs or wants and then they design a product to meet consumer preferences.

You Could Aim at People with **Specific Dietary Needs...**

Everyone should eat a healthy, balanced diet with a little bit of everything — but some people have special requirements. Your target group (see page 2) could be people with particular dietary needs:

1) Babies and toddlers need certain nutrients for growth and development.
2) Pregnant and breastfeeding women need extra protein, calcium and iron.
3) Elderly people may need to cut down on fats and carbohydrates.
4) Athletes and people with active jobs want food that provides energy.
5) Overweight people and people with inactive jobs need to eat low-fat foods.
6) Other groups, e.g. vegetarians, diabetics and people with allergies (see page 40) also have special requirements.

...Economic Needs...

EXAM TIP
You might have to explain how you'd adapt a product to make it suitable for particular groups.

How much time and money people have influences what they buy...

1) Special offers on products attract customers who want to save money.
2) But some people only buy high-quality food, never mind how expensive it is.

...Social Needs...

1) Entertaining foods are popular with children, e.g. pasta in funny shapes, cereals with free toys.
2) Trendy foods, like sushi, can be popular. Celebrity chefs can help to boost the sales of particular products, brands or supermarkets they endorse.
3) Eating can be a social occasion — people eat out at restaurants, have family Sunday roasts or eat nibbles at a party.
4) Office workers in cities tend to nip out for a sandwich at lunchtimes.
5) Convenience foods are popular with people who lead busy lives. E.g. cereal bars can be eaten on the go, ready meals can be cooked quickly, and fast food is, well, fast.

...or **Ethical Preferences**

1) People choose to buy free-range products, like eggs and chicken, because they know the animals are treated ethically.
2) Some people prefer to buy organic foods that are grown naturally.
3) Fair trade products, e.g. bananas, are popular with customers who want to make sure farmers get a fair price for their products.
4) Some people prefer to buy British or local produce, e.g. meat, to support the local economy and to reduce food miles (see page 78).
5) Some people won't eat fish that's becoming endangered, e.g. bluefin tuna.

See page 79 for more on this.

See the next page for more target groups.

Food products are designed with specific groups of people in mind

It's a bit strange really. You think you're buying something of your own free will, but really there's a manufacturer somewhere who knows you'll probably go for the low-fat, organic crisps...

Social Issues

Here are some more things you might want to think about...

*You Can Aim at People with **Cultural** or **Religious Needs**...*

You might want to consider catering for different cultural or religious needs when picking your target group.

1) You could base a product on a traditional recipe from a target culture, e.g. an Indian samosa, a Sri Lankan fish curry, a Polish chicken soup.

2) You can cater for people who obey religious food laws by using particular ingredients. In both Islam and Judaism, for example, some foods are banned (e.g. pork) and some foods must be prepared in a particular way. Food suitable for Muslims is known as halal and food that Jews can eat is kosher. So for example you could buy meat from a halal butcher.

3) Cultural and religious festivals are a good opportunity for designers and manufacturers to make special products, e.g. mince pies at Christmas and pancakes on Shrove Tuesday.

*...or just Give People **What They Want***

Not all target groups are to do with specific needs — personal taste can be just as important. People might like a product because:

1) It looks good.
2) It tastes good.
3) The packaging is appealing.
4) The food is terrifically hot and spicy.
5) Some people will try anything that's new and exotic.

*We've got **More Variety** of Food Than Ever*

Multicultural factors have a lot of influence on food production because they increase the variety of food:

1) You can eat food from all around the world, e.g. Chinese, Middle-Eastern, Italian food. Shops in multicultural areas tend to sell a wider variety of ingredients.
2) You get to try new flavours and spices, e.g. hot and spicy Mexican food.
3) Different cultures bring different cooking methods, e.g. stir fry, flambé (pouring brandy over something and setting fire to it).

Specific needs are important but so are likes and dislikes

Social factors can have a huge effect on the types of food that people buy. Manufacturers understand this and use it to help them to design food products. Make sure you are aware of the issues when you're doing your own project.

Changing Trends

You might have thought that trends only affect things like fashion, but the food you eat is influenced too.

Globalisation has Affected the Food We Eat

1) Up until the middle of the 20th century, most food eaten in the UK was grown and produced here.
2) But as trade and transport links improved, it became common for food products to be grown, processed and sold all over the world. This is known as globalisation.
3) Globalisation means that ingredients from far-away countries are now more easily available, e.g. lemongrass and exotic fruits like pineapple and mango.
4) Good transport links also mean that people can travel more, and experience different cultures for themselves, as well as allowing people from other cultures to come to the UK. This increases our exposure to multicultural foods, so there's a greater demand for them.
5) It's now common for foods to be grown and processed in countries where labour is cheaper, before being sold worldwide. This makes the final product cheaper to produce.

Tastes Change According to Trends

Manufacturers need to adapt to new trends and technologies so that people keep buying their products.

1) After the Second World War, many people emigrated to the UK, bringing their own cultures and foods with them. This made international and multicultural foods (like curry) much more popular.
2) During the 1980s, fridges, freezers and microwaves became more common. Manufacturers quickly developed new products to take advantage of these technologies — e.g. Birds Eye produced frozen foods, and microwavable meals became popular.
3) As people's lifestyles got busier, the demand for quick-to-prepare foods increased. Convenience foods and fast food became popular — e.g. the first McDonald's opened in the UK in 1974.
4) From about 1990 onwards, people have become more concerned with healthy eating. People are also paying more attention to where products come from — so they may buy local (see page 78), fair trade (see page 79) or organic foods (see page 79).

The Media Affect our Choices Too

The media influence what we buy, whether it's through clever advertising or what's reported in the news.

Advertising tries to encourage us to buy products, and can affect food trends too. TV adverts are usually scheduled to be on at times when their target group will be most likely to watch, e.g. children's snacks are often advertised in cartoon breaks (these can be high in salt and fat).

Food scares — where certain foods become linked to health problems, e.g. BSE and vCJD:

1) In the 1980s and 90s, lots of beef cows in the UK became infected with BSE (or mad-cow disease). Around the same time, some people were diagnosed with (and died from) a new human brain disease called vCJD. The two diseases are very similar, and it's thought that you could contract vCJD from eating BSE-contaminated meat.
2) When this news came out it caused a food scare, and beef sales fell dramatically — many people stopped eating British beef. Once the scare died down and controls had been put in place (to stop the disease getting in the food chain), people started eating British beef again.
3) Some people think scares are often promoted by the media, who provide shock headlines that make people buy the paper or watch the news.

Changes in transport and technology have affected what we eat

Some people think that advertising unhealthy food during children's programmes should be banned as it encourages children to eat foods that are bad for them, when they should be encouraged to eat healthily.

Environmental and Ethical Issues

Producing food, making packaging and transporting products all have a big effect on the environment. And with a growing world population, more food needs to be produced...

*Food Production can **Harm** the **Environment***

RESOURCES ARE RUNNING OUT...

1) Some food resources are in short supply. For example, stocks of many popular fish are getting very low, e.g. cod, bluefin tuna.
2) Processing food uses lots of energy, which uses up resources like oil and gas.
3) Product packaging uses up resources, e.g. trees for paper, oil for plastic, metal ores for cans.

...SO WE NEED TO USE RESOURCES SUSTAINABLY

Look in the glossary for an explanation of 'sustainable.'

1) Scarce food resources, e.g. cod stocks, need to be protected.
2) Electricity produced using renewable energy, e.g. solar power, could be used for processing.
3) Less packaging could be used or packaging could be made from renewable resources. E.g. the wood pulp used to make cardboard often comes from plantations where enough trees are planted to replace those that are felled.
4) Packaging can often be reused and recycled, instead of throwing it away in landfill sites.

TRANSPORTING FOOD HARMS THE ENVIRONMENT...

1) Some food is transported a long way to be sold, e.g. some green beans you buy in the UK have come from Kenya. This can be expensive, and it's also bad for the environment. Planes, ships and trucks all burn scarce fossil fuels and release carbon dioxide into the atmosphere, contributing to global warming.
2) But consumers want food to be available all year round, not just when it's in season here. So shops and manufacturers buy food from abroad when it's out of season at home, e.g. asparagus has a very short season here. Also, some things just can't be grown here, like bananas.
3) Transport costs (and environmental impact) can be kept down by using packaging that stacks well — to fit as much as possible on each lorry.

EXAM TIP
With questions on this kind of stuff, don't write things like "because it's bad for the environment" — explain why it's bad (or good...).

...SO LOCAL AND SEASONAL FOOD IS BEST

1) To reduce food miles, some people try to only buy local products.
2) So, if you're developing a fruit tart, consider whether the fruits you intend to use are available locally and whether they're in season.

Food miles is the distance food travels from where it's produced to where it's sold.

Environmental and Ethical Issues

Some Food Meets **Environmental** and **Ethical Standards**

Many consumers care about different environmental, ethical and social issues. So some companies develop products that meet specific standards — and they put labels on the products to show this.

Free range foods

1) Food labelled as free range lets consumers know that animals have a higher standard of welfare than in intensive farming and they're free to roam.
2) They cost more because it's a less efficient way of farming.

Farm Assured foods

1) The Red Tractor symbol lets consumers know that the food producers meet standards of food safety, hygiene, animal welfare and environmental protection set by the Assured Food Standards scheme. (There are other similar farm assured schemes.)
2) Farm assured foods can be traced back to the farms they come from.

Organic foods

1) Food labelled as organic is grown without using any artificial pesticides or fertilisers.
2) Organic meat production has really high animal welfare standards and the animals aren't given growth hormones.
3) However, organic food production isn't as efficient — less food is produced per acre.
4) This makes it more expensive but some people are willing to pay more for food that's grown naturally.

Fair trade foods

1) The fair trade movement tries to make sure that workers and farmers in developing countries are paid fairly and have good working conditions. There's a minimum price for fair trade produce, so it's usually more expensive (i.e. if the market value of a product drops, fair trade farmers aren't affected).
2) Through fair trade, workers can invest in their communities, e.g. to build schools or health centres.
3) The only downside to the scheme is that fair trade producers often produce a lot because of the good prices — and they sometimes produce too much. This can make world prices fall and cause producers who aren't in a fair trade scheme to lose out.

Fair Trade Certification

The FAIRTRADE Mark is used on products that meet international Fair trade standards, e.g. bananas, cocoa. It's the consumer's guarantee that producers have been paid an agreed and stable price which covers the cost of sustainable production.

Look at the label to check if food meets environmental and ethical standards

There are loads of ways to reduce the impact of food production on the environment — so make sure you know lots of examples. Learn the key terms like sustainability too — they could get you easy marks in the exam.

The 6Rs

The 6Rs crop up everywhere — they're important when you're thinking about sustainability or diet and nutrition. Learn them well and have them at the back of your mind whenever you're designing a product.

You Should Learn the **6Rs**

The 6Rs are: Recycle, Reuse, Reduce, Refuse, Rethink and Repair.

Using the 6Rs can help reduce the environmental impact of your lifestyle, as well as making it healthier.

They Can Help You **Design** a **Sustainable Product**

RECYCLE

1) Recycling means we reuse old resources instead of using up new ones.
2) Packaging can often be recycled — steel/aluminium, plastic, glass, card and paper are all recyclable. Some products have packaging made from recycled materials — e.g. a cereal box could be made from recycled cardboard.

Although it saves resources, recycling sometimes takes more energy than making new products. It can also be expensive.

3) Some types of packaging are biodegradable (they naturally rot in the environment) — so they won't add to landfill.
4) Some food waste (often veg scraps) can be composted.

REUSE

1) Old products (or parts of them) can be used again for the same or a different purpose, e.g. glass or plastic bottles and containers can be reused at home. This stops more resources being used up.
2) Manufacturers can reuse leftover food, e.g. when sugar is made, the sugar beet waste can be used to feed pigs.
3) You can reuse leftovers at home too — e.g. you can use stale bread to make bread-and-butter pudding or bread crumbs.

REFUSE

1) Refuse to use packaging or ingredients that are unsustainable (see page 78) — find alternatives.
2) Refuse to buy products with excess packaging.

The 6Rs

REDUCE

1) Reduce the amount of resources you use when you make your product — like energy, ingredients, materials, chemicals (e.g. pesticides) and packaging.
2) Think about ways to reduce the energy used to cook or transport your product, e.g. changing the shape of packaging so lots can be transported together.
3) You can reduce waste by recycling and reusing things. Avoiding buying too much food helps you cut down on waste as well.

REPAIR

1) Lots of large, expensive equipment is used in food production. It should be kept in good condition and repaired when it breaks down so resources aren't used up making more.
2) Try and repair your home equipment before buying more. You can buy replacement parts for some things.

RETHINK

Don't get bogged down by traditional ideas — rethink how old products could be made in sustainable ways. E.g. reducing the packaging around an Easter egg.

*Four of the **6Rs** Apply to **Nutrition** as well*

Refuse, rethink, repair and reduce can help you improve your health — or the nutritional value of a product you're designing. Here's how:

REFUSE

Refuse to eat unhealthy products — products don't need to be high in fat, salt or sugar to taste nice.

RETHINK

1) Rethink how products could be made with lower fat, salt and sugar but still taste great.
2) Re-market 'healthy' products so they're more appealing.

REPAIR

Don't just think about low fat, sugar and salt. Think about how other nutrients help to repair and maintain a healthy body. Use these in your product, e.g. added vitamin C or folic acid.

REDUCE

1) You can reduce the amount of sugar, salt and fat you use — both in your diet and in your product.
2) Reducing the amount of processed foods you buy will help you do this.

EXAM TIP
The 6Rs crop up everywhere — make sure you can use them in relation to sustainability and nutrition.

Recycle, reduce, revise...

There's lots to learn on these two pages — just remembering all six Rs can be tricky. Cover the page and check that you've learnt them all. Make sure you can relate them to your diet, as well as sustainability...

Labelling

Labelling on products can help people make informed choices about what they eat. Manufacturers also use labels to try to tempt customers with slogans like 'a healthy choice' and 'good value for money'.

Food Labels **Can't be Misleading**

Manufacturers can't just say anything they like on the label — they must obey these laws:

Trade Descriptions Acts (1968)

Food Labelling Regulations (1996)

Food Safety Act (1990)

Food Standards Act (1999)

Labels **Must** Tell You **Certain Information** by **Law**

The law says that the label on pre-packed food has to tell you all this stuff:

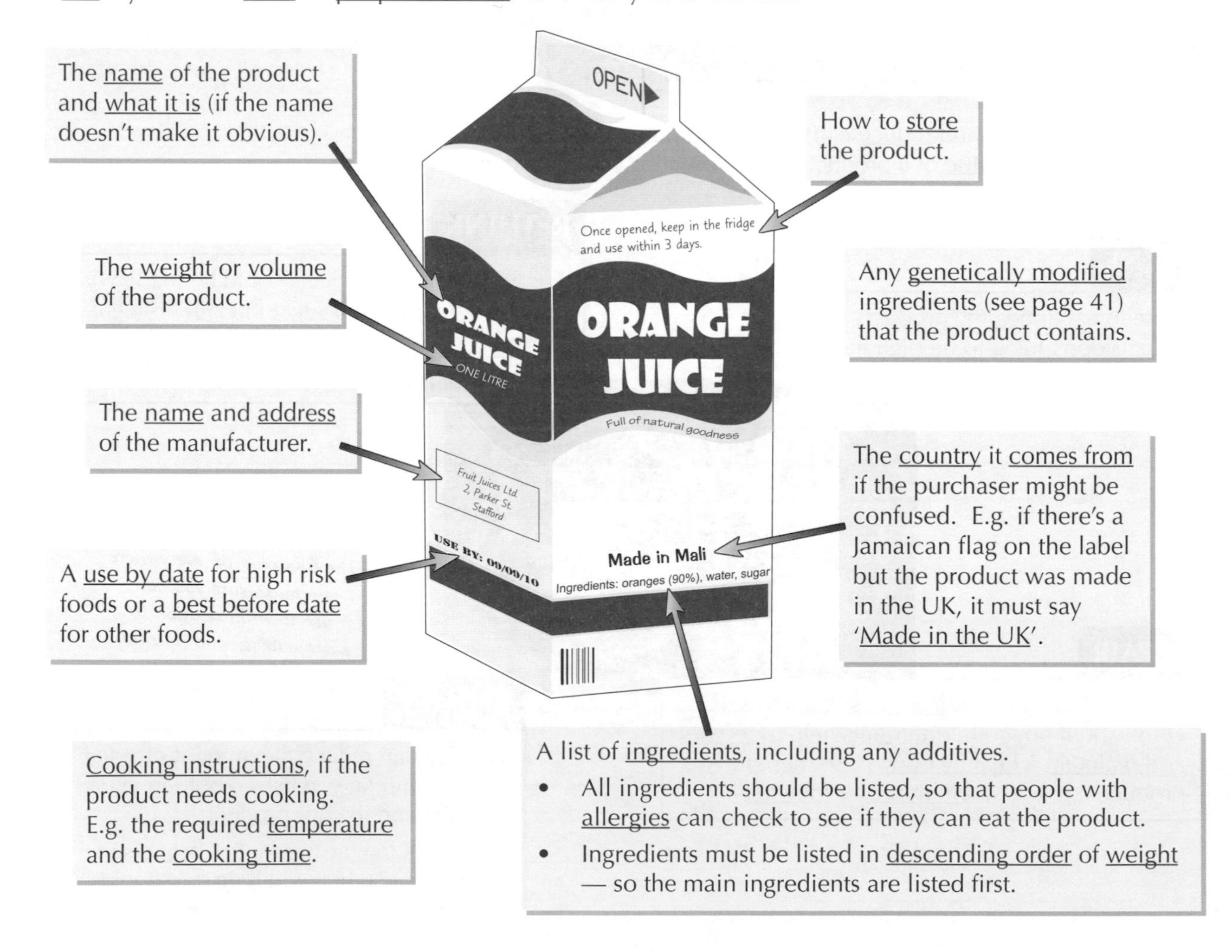

Learn what the labelling on food packaging must include

So, only pre-packed foods need to be labelled by law. Simple. (Or is it? — a single cucumber wrapped in plastic film doesn't count as pre-packed. Hmm, I'd just say 'pre-packed' if it comes up in the exam.)

Labelling

There's often loads of extra information on food labels — you need to know this stuff too.

Nutritional Information is Needed to Back up Claims

1) Lots of products list nutritional information but they don't have to by law...
2) ...UNLESS they make a special nutritional claim, such as 'low-fat' or 'high in fibre'.
3) Nutritional information is usually shown in a table listing energy content, protein, carbohydrate, etc.
4) Claims such as 'low fat' can only be made if the nutritional information backs this up.

NUTRITIONAL INFORMATION	per 100g	per 55g serving
Energy	2180kJ/525 kcal	1199kJ/289 kcal
Protein	6.5g	3.6g
Carbohydrate	50.0g	27.5g
of which sugars	2.0g	1.1g
Fat	33.0g	18.2g
of which saturates	15.0g	8.3g
Sodium	0.7g	0.4g
Fibre	4.0g	2.2g

1 kcal is 1 calorie.

There can be Other Useful Stuff on Labels too

Other information doesn't have to be there but manufacturers try to make their label useful to consumers.

Some products guarantee the food is of a high standard — or else you can claim your money back.

Symbols are used to show that food is suitable for a particular diet, e.g. food suitable for vegetarians is often shown with a green V.

The manufacturer can suggest accompaniments to the product — what kind of food it's best eaten with, e.g. a label on chutney may say it's best eaten with cold meats or cheese.

Possible allergy problems can be highlighted, e.g. 'may contain traces of nuts'.

Traffic-light labelling on a product shows how healthy it is at a glance. Red, orange and green colours show whether a product has high, medium or low amounts of saturated fat, salt and sugar. For example, a pizza might be red for saturated fat and yellow for salt and sugar.

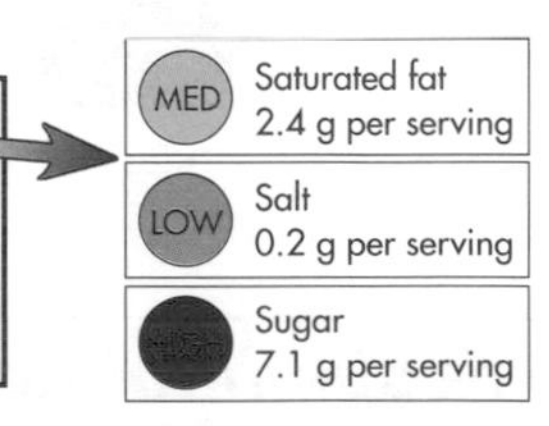

A recycling symbol means that some (or all) of the packaging can be recycled.

Some labels on take-away products warn about very hot contents.

Traffic-light labelling can make nutritional information easier to understand

Make sure you know what all the labelling on food packaging means — and learn whether it has to be there by law or not. Understanding how to read package labelling is useful for your exams and for supermarket trips.

Packaging

Packaging is pretty useful — you wouldn't want to buy your food and then have it all slopping about and mixing together in your shopping bags... Even better, some packaging can stop food going off.

Packaging **Contains**, **Protects** and **Preserves** Food

Most food products are packaged before they're sold:

1) To contain the product neatly.
2) To protect it from being damaged while it's being transported, displayed and stored.
3) To preserve the food and extend its shelf life — otherwise it's more likely to be wasted.
4) To avoid contamination, e.g. from flies, vermin or people touching the food.
5) To identify what the product is and to give customers useful information.

There are laws about food packaging:

1) It can't be hazardous to human health.
2) It can't cause food to deteriorate (go off).
3) It can't cause an unacceptable change in a product's quality.

Different Forms of Packaging can **Extend Shelf Life**

1 MODIFIED ATMOSPHERE PACKAGING (MAP)

MAP extends the shelf life of fresh foods, e.g. fresh and cooked meats, fresh pasta, cheese and sandwiches.

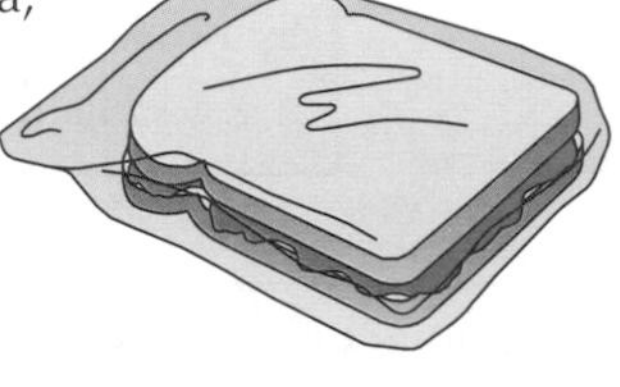

1) The food is put into plastic packaging with a mixture of oxygen, nitrogen and carbon dioxide in particular proportions. It's then sealed and chilled.
2) But once the packet's been opened, the food has a normal shelf life.

2 VACUUM PACKAGING

Vacuum packaging is often used for dry foods, e.g. coffee, and for meat and fish.

1) Food is put into plastic packaging, then the air is sucked away from around the food. It's then sealed to keep the food in oxygen-free conditions.
2) Once the packet is open you have to follow the storage instructions.

Nanotechnology can **Improve Packaging Properties**

Nanotechnology is a new technology that involves using very, very small particles (nanoparticles).

1) Some nanoparticles can make packaging stronger, lighter or more heat-resistant.
2) Food can be made to last longer, e.g. adding clay nanoparticles to plastic makes the packaging better at keeping out oxygen and moisture. Some nanoparticles can kill harmful microorganisms.
3) Some 'smart packaging' uses nanoparticles to change the packaging's properties depending on the conditions. E.g. a milk carton could be made to change colour when the milk goes off.

Food packaging can make food stay fresh for much longer than normal

You may think that a MAP is something you take on a long walk, but no — it's a type of packaging too. You've probably seen it loads of times in the shops, but now you need to know what it is and how it keeps your food fresh.

Packaging

Loads of different materials can be used to package food, but deciding what type to use isn't always straightforward. It's not just about a pretty packet — all packaging materials have advantages and disadvantages.

Different Types of Material are Used for Packaging

1) Packaging uses up a lot of resources — some of which are finite (will run out eventually).
2) Manufacturing packaging uses a lot of energy.
3) However, using packaging does mean that less food is wasted — so there's a balance.
4) Using recycled material to make packaging reduces its environmental impact.

Various materials are used in different shapes and thickness to make packaging for different products.

GLASS, e.g. bottles, jars

- It's a strong, rigid material
- It's transparent — customers can see what they're buying
- It's resistant to high temperatures
- It can be reused and is easy and cheap to recycle

BUT...

- It's pretty heavy
- Glass breaks easily

PLASTIC, e.g. bottles, trays

- You can get rigid plastics and flexible ones
- It can be transparent or coloured
- Many types are microwavable — food can be heated in the packaging
- It's lightweight
- It can be printed on

- Most types don't biodegrade
- Some plastic can't be recycled

CARD and PAPERBOARD, e.g. boxes, packets

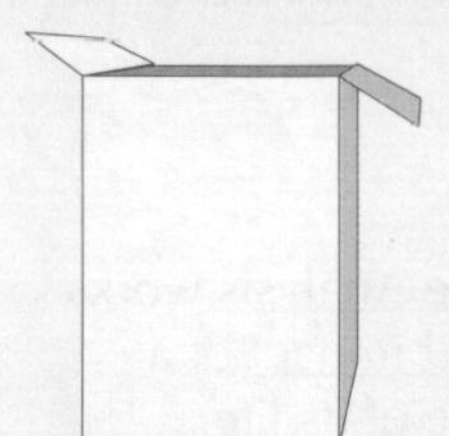

- Usually biodegradable
- Fairly strong
- Lightweight and flexible
- Easy to print on
- Waterproof if laminated
- Easy and cheap to recycle

BUT...

- You can't see the contents, and it's not very rigid, so the product may get squashed

METALS (aluminium, tin, steel), e.g. cans, foil

- Most metals are strong and some are fairly light e.g. aluminium
- They're resistant to high temperatures
- Aluminium is cheaper to recycle than to extract from the ground

BUT...

- Metals can react with some foods
- You can't see the contents

There's no such thing as perfect packaging material

They're all good in some ways but not so good in others. Try covering the page and making a table of the different materials, their advantages and their disadvantages — make sure you can remember them all.

Packaging and the Environment

Packaging may be great for stopping food from going off quickly, but it's not so good for the environment.

*Packaging can be **Bad** for the **Environment***

1) Manufacturing packaging uses a lot of energy and resources — some of which are non-renewable (e.g. oil products are used to make plastic containers).
2) Most products need to be packaged to stop the food from getting damaged and wasted.
3) Some products have excess packaging just to make them look more attractive on the shelf — e.g. products that are shrink-wrapped within a box or other container.
4) Packaging often gets used once, thrown away and then just takes up space in Britain's already huge landfill sites. Not only is this a waste of the materials, but some packaging, like plastics, take a long time to biodegrade, and could take up space in a landfill site for years.
5) Packaging also adds to the weight of a product, so transporting it uses more energy. This uses up more fossil fuels and produces more greenhouse gases, which add to global warming.

*You can **Reduce** the **Environmental Impact***

There are lots of ways to reduce the environmental impact of packaging:

1) You can recycle tins, plastic, glass, card and paper — look out for this symbol:
2) You can buy products with little or no packaging — or refuse to buy products with excess packaging. This is even better than recycling it. Also, products with less packaging are often cheaper, as manufacturers don't have to spend money on it.
3) You can choose products with biodegradable packaging.
4) You can choose products with packaging made from recycled materials — they may not look as pretty, but they're better for the environment.
5) You can carry your food in reusable shopping bags — this reduces the need for plastic bags, which end up in landfill.

EXAM TIP
If you get a question on packaging in your exam, make sure you think about the 6Rs (see page 80).

In 1997, the Government made a set of rules for all businesses that manufacture, fill or sell packaging. The point was to:

1) Increase the amount of packaging that can be recycled.
2) Reduce the amount of packaging in total.

*You Can Measure **Environmental Impact***

Your Carbon Footprint measures the impact your lifestyle has on the environment — in terms of the amount of greenhouse gases produced (especially carbon dioxide). Burning fossil fuels for heat, electricity, transport etc. increases your carbon footprint. Foods have a carbon footprint too — and the further a product has to travel, the larger its carbon footprint.

A product's Life-Cycle Analysis works out the environmental impact of a product at every stage of its life. It goes from sourcing the materials to disposing of the waste produced (and every step on the way, e.g. transport).

Your Eco Footprint is a bit like a carbon footprint, but it measures the way you use the planet's natural resources and how much waste you produce, and balances it against how well the Earth can produce new resources and absorb your waste.

Biodegradable and recycled packaging is better for the environment

This is the last revision page of the book — it snuck up very quietly. My work isn't done yet though, there's plenty of exam-style questions on the next few pages to get you ready for the real thing. Good luck.

Warm-Up and Worked Exam Questions

Doing these questions will soon let you know if you've got the basic facts straight. If you don't know them, you'll struggle in the exam — so take the time to go back over the bits you don't know.

Warm-up Questions

1) Amy is designing a range of food products targeted at rich people who work long hours.
   a) Suggest one product she could include in the range.
   b) Explain why your product would be suitable.
2) What is the fair trade movement? Give one advantage and one disadvantage of fair trade.
3) Give an example of how leftover food can be reused:
   a) at home b) in the food industry
4) Emma is setting up a business selling fresh pasta. She is considering what type of packaging to use.
   a) Suggest a packaging technique she could use to keep the pasta fresh.
   b) Outline the process used to package food this way.

Worked Exam Questions

I'm afraid this helpful blue writing won't be there in the exam — so make the most of it now.

1 Snazzy Sarnies is a food company that is aiming to make its manufacturing processes more sustainable. One of their products is shown below.

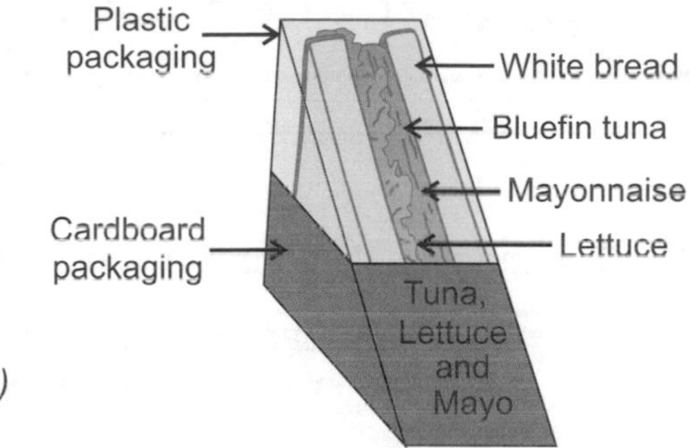

(a) (i) Explain what is meant by the term 'sustainable'.

A process or material that can be used without causing permanent damage to the environment or using up finite resources.

(1 mark)

(ii) Give two reasons why the product shown on the right might not be sustainable.

It contains bluefin tuna which is in short supply. Energy is used to process the food, using up resources like oil and gas.

Think about the sandwich ingredients as well as the type and amount of packaging.

(2 marks)

(b) (i) One way Snazzy Sarnies could make more sustainable products is to reduce the food miles of their ingredients. Explain what is meant by the term 'food miles'.

The distance food travels from where it's produced to where it's sold.

(1 mark)

(ii) Give one advantage to the company, other than improving sustainability, of reducing the food miles of their ingredients.

It lowers the cost of transporting the food.

(1 mark)

(c) Suggest two other ways that Snazzy Sarnies could make their products more sustainable.

It could reduce the amount of packaging. It could only use packaging that is made from renewable resources, e.g. cardboard.

(2 marks)

Exam Questions

1 Products are often targeted at groups of people with special dietary needs. For each product listed below, name one possible target group and explain how it meets their dietary needs.

Product	Target Group	Reason
Fruit salad	....	....
High energy snack bar	....	....
Calcium-enriched cereal	....	....

(6 marks)

2 The nutritional information below is from the label of a snack product.

(a) State how many grams of fat are found in 100 g of the snack product.

....

(1 mark)

(b) The manufacturer decides to label the snack as 'low-fat'. State whether this is a reasonable claim and explain your answer.

NUTRITIONAL INFORMATION		
	per 100 g	per 40 g serving
Energy	2180kJ/525 kcal	872kJ/210 kcal
Protein	6.5 g	2.6 g
Carbohydrate	50.0 g	20.0 g
of which sugars	2.0 g	0.8 g
Fat	33.0 g	13.2 g
of which saturates	15.0 g	6.0 g
Sodium	0.7 g	0.3 g
Fibre	4.0 g	1.6 g

....

....

(2 marks)

3 Food is often packaged before it is sold. Give three functions of packaging.

1. ....

2. ....

3. ....

(3 marks)

Revision Summary for Section Four

This might not be the most interesting page in the section, but it's a useful one. Have a go at answering these questions, looking up the bits you don't know. Then go through them once more without looking bits up.

1) Briefly describe the dietary needs of:
   a) babies b) pregnant women c) athletes
2) Give three examples of how ethical preferences affect what foods people buy.
3) Give two reasons why designers might want to know the religions of people in their target groups.
4) Explain what globalisation is, and how it affects the food we eat.
5) Explain how food manufacturers adapted to the inventions of freezers and microwaves.
6) Describe what is meant by a 'food scare'.
7) Give an example of a food resource that is in short supply.
8) a) Suggest a renewable energy source that can be used to generate the electricity needed for processing food.
   b) Suggest a renewable resource that can be used to make food packaging.
9) What does 'free range' mean?
10) How does fair trade help communities?
11) a) What does 'organic' mean?
    b) Why is organic food more expensive than food produced by intensive farming?
12) What are the 6Rs?
13) Name three materials used for packaging that can be recycled.
14) What are 'biodegradable materials'? Why are they better for the environment?
15) Which four of the 6Rs relate to diet and nutrition?
16) What nutrients should you try to reduce in your diet?
17) Name two vitamins that can be used to help repair the body.
18) What kinds of food product must be labelled by law?
19) a) Why is important that all of the ingredients are listed on a food label?
    b) Give five things, other than the ingredients, that the label must include.
20) Under what circumstances do manufacturers have to give nutritional information on their products?
21) Many food labels include things that don't need to be there by law, e.g. 'traffic-light labelling'.
    a) What is traffic-light labelling?
    b) Give two other things that a food label might have, that don't have to be there by law.
22) Give two ways in which nanotechnology can be used in packaging to make food last longer.
23) How does vacuum packaging extend the shelf life of food?
24) a) Give two advantages and two disadvantages of using glass for food packaging.
    b) Give two advantages and two disadvantages of using metals for food packaging.
25) Charlotte is making a microwavable meal that can be cooked in its packaging.
    What type of material should she use for her packaging?
26) Give three ways that packaging can be bad for the environment.
27) Give two ways that you can reduce the environmental impact of packaging.
28) What is meant by the 'carbon footprint' of a food?
29) What is a product's 'Life-Cycle Analysis'?

Glossary

6Rs	Recycle, Reuse, Reduce, Refuse, Rethink, Repair.
acetic acid	The acid in vinegar.
additive	Something that's added to a food product to improve its properties.
aerate	To add air to a mixture to help make it lighter, e.g. when making cakes.
aesthetics	How a product looks.
alternative protein	A form of protein other than protein from meat (e.g. tofu, TVP), which is suitable for vegetarians.
amino acids	Proteins are made of amino acids.
ascorbic acid	Another name for vitamin C.
balanced diet	A healthy diet that contains a bit of everything your body needs.
batch production	Making a certain number of a product in one go.
binding	Holding ingredients together so the product doesn't fall apart. For example eggs are used to bind ingredients in burgers.
biodegradable	A biodegradable material is something that rots down naturally.
CAD	Computer-aided design.
CAM	Computer-aided manufacture.
caramelisation	When sugar is heated and forms a sweet-tasting, brownish liquid.
carbon footprint	A measure of the impact something has on the environment, based on the harmful greenhouse gases produced.
citric acid	The acid in lemon juice.
closed question	A question with a limited number of possible answers, e.g. do you like spicy food?
coagulate	To change into a more solid state. For example if you fry an egg it coagulates.
conduction	The transfer of energy through solids.
consistent	The same every time.
contaminate	To make something dirty and unhygienic, e.g. a fly could contaminate your soup.
continuous flow	Continuous flow production means making large numbers of a product non-stop.
control point	A stage in a process where you put in a control to stop a problem occurring.

Glossary

convection	The transfer of energy through gases (e.g. air) or liquids.
cross-contamination	Transferring potentially harmful bacteria from one thing to another, e.g. via work surfaces, equipment or your hands.
danger zone	The range of temperatures (5 °C to 63 °C) in which bacteria multiply very quickly.
deficiency	Not getting enough of something, e.g. calcium deficiency.
design brief	A short statement explaining why there's a need for a new product.
design criteria	The general characteristics a product should have, e.g. "Appealing to children." A list of design criteria is sometimes called a design specification.
deteriorate	When the quality of food decreases or it 'goes off'.
disassembly	Taking a product apart.
dormant	Inactive, a bit like a deep sleep — bacteria become dormant in frozen food.
E number	A number (e.g. E150a) given to an additive when it passes EU safety tests. The additive can then be used in food throughout the European Union.
eatwell plate	Government healthy eating guidelines in the form of a pie chart which shows how much or little of each food group your diet should contain.
eco footprint	A measure of the human demands on environmental resources.
emulsifier	Something that keeps an oily and watery mixture stable (stops it separating into two layers).
emulsion	A mixture of oily and watery liquids, e.g. salad dressing.
enriching	Adding something like butter or cream to a product to make it thicker and tastier.
enrobing	Coating a food product in something, e.g. a thin layer of chocolate.
Environmental Health Officer (EHO)	A person who monitors public and environmental health standards, e.g. by doing routine hygiene inspections.
enzymic browning	The reaction that happens when fruit is sliced open and left uncovered — the surfaces of the cut pieces turn brown.
essential amino acids	Amino acids that the body can't make itself so you need to get them from your diet.
ethical issue	A moral issue — when many people have views about whether something's morally right or wrong.

Glossary

fair trade	When workers in developing countries get a fair price for their food and have good working conditions.
feedback	Sending back information, often so that a person or a computer can monitor whether a process is working as it should.
fermentation	When yeast breaks down sugars to release carbon dioxide and alcohol.
finishing techniques	Techniques that are used to make the finished product look as good as possible, e.g. glazing, icing.
five a day	The Government recommends that everyone should eat at least five portions of different fruits or vegetables every day in order to be healthy.
food miles	The distance a product travels from where it's produced to where it's sold.
food scares	When a particular food is linked to a health problem — often people stop buying it.
fortification	When extra nutrients are added to food or drink.
fossil fuels	Coal, oil and natural gas, or fuels made from them. e.g. petrol. Burning fossil fuels, e.g. for transport, releases carbon dioxide, which contributes to global warming.
free-range	Free-range animals have more space to live — they're free to roam.
functional food	A food that has been artificially modified to provide a particular health benefit on top of its normal nutritional value.
Gantt chart	A time plan that shows how long different tasks will take and the order they need to be done in.
gel	The semi-solid structure you get when a small amount of a solid ingredient sets a lot of liquid, e.g. jelly.
gelatinisation	When starch particles swell and burst, thickening a liquid.
gelling agent	Something that causes a liquid to thicken and set as a gel.
genes	The 'instructions' for how to develop contained in all the cells of a plant or animal. Genes control the characteristics of the plant or animal, e.g. how quickly fruit ripens.
genetically modified (GM) food	Food that's had its genes altered to give it useful characteristics. For example, GM tomatoes with a longer shelf life than normal.
glazing	Adding a coating to give a product a shiny, glossy appearance.
globalisation	When products grown or produced in one country, are processed and sold all over the world.
gluten	A protein formed when dough is kneaded, that makes dough stretchy.

Glossary

Guideline Daily Amounts (GDAs)	Information about how much energy or how much of certain nutrients an average adult needs each day.
HACCP	Hazard Analysis Critical Control Points.
hazard	Anything that could go wrong or cause harm.
heat transference	When heat energy moves from one place to another — by convection, conduction or radiation.
high-risk food	A food in which bacteria can grow quickly.
hydrogenation	A process that makes oils more solid at room temperature, e.g. to make margarine.
landfill	A landfill site is a large rubbish dump that's eventually covered over with earth.
lecithin	A natural emulsifier found in egg yolks.
life-cycle analysis	A way of working out a product's environmental impacts.
local produce	Food that is produced locally, so doesn't have to be transported as far. Often found at farmers' markets.
manufacturer's specification	Precise instructions that tell the manufacturer exactly how to make a product.
marinate	To soak something in a mixture of things before cooking to give it more flavour, e.g. oil, wine, vinegar and herbs.
mass production	Making large numbers of a product, often on an assembly line or conveyor belt.
model	A test version of a product that you make during the development stage.
modified starches	Starches that have been treated so that they react in a particular way in certain conditions, e.g. they're used in packet custard that thickens instantly. (Modified starches are also called smart starches.)
monosodium glutamate (MSG)	A natural flavour enhancer which boosts the existing flavour of a product and gives it a savoury taste.
nanoparticles	Very, very, very small particles of a substance. Nanoparticles of a substance often have different properties from the 'normal' substance.
nanotechnology	A new technology that involves using nanoparticles.
non-starch polysaccharide (NSP)	Often called dietary fibre. It's a type of carbohydrate that isn't digested by your body.

Glossary

nutrients	Proteins, carbohydrates, fats, vitamins and minerals are all nutrients.
nutrient deficiency	Not getting enough of a nutrient, e.g. calcium deficiency.
nutrient excess	Getting too much of a nutrient, e.g. salt excess.
one-off production	Making single products that are unique.
open question	A question that has no set answers, e.g. why don't you like puddings?
organic	Organic crops are grown without using any artificial pesticides or fertilisers. Organic meat is produced to very high welfare standards and without artificial growth hormones or the regular use of antibiotics.
palatability	How pleasant the taste of a food is.
pectin	A natural gelling agent found in some fruits.
pesticide	A chemical or other substance used to kill pests.
preservative	Something added to food to slow the growth of bacteria so that food lasts longer.
preserve	To make food last longer.
processed foods	Foods that have been processed in some way, usually for our convenience, e.g. tinned meats. They often have extra salt added so they can be unhealthy.
product specification	A detailed description of how the product should look (including measurements) and taste — it also includes what ingredients will be used.
prototype	A full-size, one-off model of a design. A prototype is made so that you can check you're completely happy with the product before making lots of it.
quality control	Checking that the standards you've set for the quality of a product are being met.
radiation	The transfer of energy through waves of radiation.
raising agent	Something that releases bubbles of gas that expand when heated. Raising agents are used to make cake and dough mixtures rise.
Recommended Daily Amounts (RDAs)	How much of certain vitamins and minerals an average adult needs each day.
recyclable	A recyclable material is one that could be recycled fairly easily.
Red Tractor symbol	A symbol showing that food can be traced to the farm it came from and that the producers meet standards for safety, animal welfare and environmental protection.
renewable	A renewable resource is one that's replaced by natural processes as fast as it is used up by humans, e.g. softwood trees in a plantation.

Glossary

resources	Things you need to make new products, e.g. oil is a resource used to make plastic.
risk assessment	Identifying potential hazards and the precautions needed to minimise risks before work starts.
roux	A sauce base made from plain flour and melted butter.
salmonella	A bacteria that causes food poisoning. It's often found in eggs and chicken.
saturated fats	A group of fats that come mainly from animal sources and are solid or semi-solid at room temperature. Eating too much of them increases your risk of heart disease.
seasonal foods	Foods that are only produced in a particular season, e.g. British-grown asparagus is only available in May and June.
sensory analysis	Tasting samples of food and rating how good they are in various ways, e.g. taste, texture. It's done to find out what consumers think about new or existing products.
shelf life	The length of time a product can last without going off or losing its quality.
shortening	The effect of adding fat to a floury mixture — giving a product a crumbly texture.
solution	What you get when a solid dissolves in a liquid.
standard food component	A ready-made ingredient or food part, e.g. a ready-made pizza base.
suspension	What you get when a solid is held in a liquid but doesn't dissolve.
sustainable	A sustainable process or material is one that can be used without causing permanent damage to the environment or using up finite resources, e.g. sustainable wood comes from forests where fast-growing trees are chopped down and replaced.
syneresis	When protein coagulates and squeezes the fat and water out of a food.
target group	The group of people you want to sell your product to.
tenderising	Making meat more tender so that it's easier to eat, e.g. by marinating it before cooking.
test kitchen	A kitchen used to develop new food products.
toxic	A toxic chemical is one that's harmful to health.
unsaturated fats	A group of fats that come mainly from vegetable sources and are usually liquid at room temperature.
viscous	A thick, syrupy consistency.
work order	A table or a flow chart that shows tasks in sequence.

Practice Exam

Once you've been through all the questions in this book, you should feel pretty confident about the exam.
As final preparation, here is a **practice exam** to really get you set for the real thing.

General Certificate of Secondary Education

GCSE
Design and Technology
Food Technology

Centre name					
Centre number					
Candidate number					

Time allowed: 2 hours

Surname
Other names
Candidate signature

In addition to this paper you will need:
- Drawing equipment
- Coloured pencils

Instructions to candidates
- Write your name and other details in the spaces provided above.
- Answer **all** questions in the space provided.
- Use blue or black ink or ball-point pen.
  You should use pencil and coloured pencils for drawing only.

Information for candidates
- The marks available are given in brackets at the end of each question or part-question.
- There are **6** questions in this paper.
- The maximum mark for this paper is **120**.

Advice to candidates
- Work steadily through the paper.
- You should use good English and clear presentation in all your answers.

Leave blank

1 A bakery is developing a new range of pastry products.

The product should:

- have a savoury filling
- appeal to vegetarian consumers
- have a glazed finish
- offer sensory appeal

(a) Use notes and/or annotated sketches to produce two different design ideas that meet the design criteria for the pastry product. Do not draw any packaging.

(2 x 6 marks)

Turn over

(b) Choose your best design idea.

Best design: ....

Using a flowchart or diagrams and notes, draw a plan for making your chosen design in a test kitchen. Include control checks in your plan.

(8 marks)

Leave blank

(c) Vitamins are essential for good health. Complete the table below to show two vitamins provided by the ingredients in your chosen product design. State the function of each vitamin in the body. An example has been done for you.

Vitamin	Ingredient	Function in the body
Vitamin B Group	Flour	They help the nervous system, the release of energy from carbohydrates and tissue growth and repair.
....	....	....
....	....	....

(6 marks)

(d) The bakery is considering using two different types of pastry in their new range of products. They carry out sensory analysis testing on the two pastries. The results are shown below.

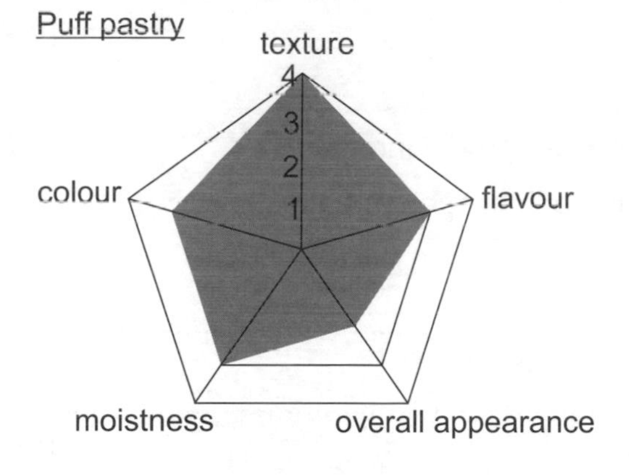

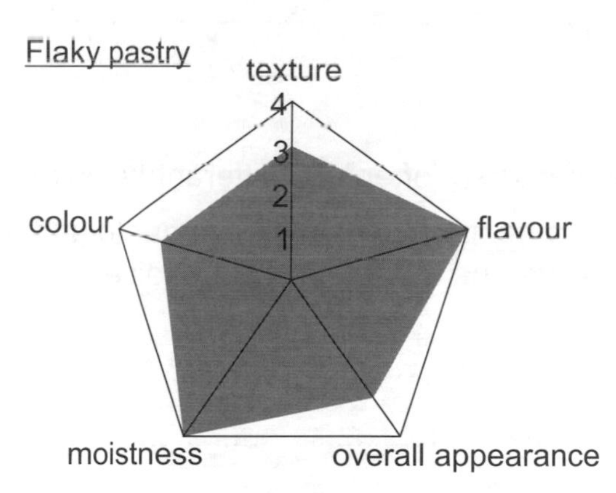

Analyse these results and suggest how you would use these findings when designing and developing your pastry product.

....

....

....

....

....

(4 marks)

30

Turn over

Leave blank

2 (a) Eggs have various functions in cooking. Complete the table below by describing the function of the egg in the different foods listed. An example has been done for you.

Food	Function of the Egg	Description
Burgers	**Binding**	Eggs coagulate and stick the ingredients together as they cook.
Quiche	**Coagulation**	....
Mayonnaise	**Emulsification**	....
Cake	**Aeration**	....

(6 marks)

(b) Manufacturers often use different types of sugar when producing cakes.

(i) Name the most suitable sugar to use to make a light sponge cake.
Give one reason for your choice.

....

....

(2 marks)

(ii) Name the most suitable sugar to use to decorate a cake.
Give one reason for your choice.

....

....

(2 marks)

Leave blank

(c) Alkalis are often used in cake mixtures to make cakes rise.

(i) Name one alkali that acts as a raising agent.

....

(1 mark)

(ii) Briefly explain how the alkali you named in part (i) acts as a raising agent.

....

....

....

(2 marks)

(iii) Give one disadvantage of using alkalis in cakes.

....

(1 mark)

(iv) Name two different products it would be suitable to make using an alkaline raising agent. Explain why the disadvantage given in (iii) would not be a problem for these products.

....

....

....

....

(3 marks)

17

3 A basic tuna pasta bake is made up of tuna, vegetables, pasta, white sauce and cheese.

(a) Explain the function of starch in each of these components:

Component	Function of Starch
Pasta	....
White sauce	....

(4 marks)

(b) Describe different ways in which the tuna pasta bake could be adapted for:

(i) a consumer who follows a vegan diet,

....

....

....

(3 marks)

(ii) a consumer with coeliac disease,

....

(1 mark)

(iii) a consumer with nut allergies.

....

(1 mark)

(c) Some people need to avoid drinking cow's milk and eating dairy products.

(i) What is the name given to this medical condition?

....

(1 mark)

(ii) Identify two ingredients in the tuna pasta bake that they would be unable to eat.

....

....

(2 marks)

(d) The symbols below are found on the labels of certain pasta bake products.
State the meaning of each symbol and give two reasons why each symbol is useful.

(i)

V

....

....

....

(ii)

May contain nuts

....

....

....

(iii)

....

....

....

(9 marks)

21

Leave blank

4 A manufacturer is designing a pizza. It will be made using dough, a tomato-based sauce, mozzarella cheese and ham.

(a) The manufacturer decides to use pre-made tomato sauce to make the pizza.

(i) Give three advantages to the manufacturer of using pre-made tomato sauce.

....

....

....

....

....

(3 marks)

(ii) The manufacturer decides that the pre-made tomato sauce lacks flavour.
Suggest two ways the flavour of the sauce could be improved.

1. ....

2. ....

(2 marks)

(iii) The pre-made tomato sauce is a standard component.
Name one other standard component that can be used in pizza production.

....

(1 mark)

(b) Explain the harmful effects that using standard components can have on the environment.

....

....

....

....

....

....

....

(4 marks)

Leave blank

(c) Sales of pizza as a take-away meal have changed in recent years.
The table below shows pizza sales in the UK from 2005 to 2009.

Year	2005	2006	2007	2008	2009
Pizza sales	£600 million	£560 million	£520 million	£700 million	£750 million

(i) Using the information in the table, describe how pizza sales have changed since 2005.

....

....

....
(2 marks)

(ii) The average consumer had less money to spend in 2009 than in 2007.
Suggest one reason for the trend in UK pizza sales between 2007 and 2009.

....

....
(1 mark)

(d) Give examples of how Computer-Aided Design (CAD) and Computer-Aided Manufacture (CAM) could be used in pizza production.

....

....

....

....

....

....
(4 marks)

17

Leave blank

5 A Chinese restaurant sells a dish containing seafood, vegetables, sweet and sour sauce and rice.

(a) Describe three different control checks the restaurant could use when selecting ingredients to produce this dish.

....

....

....

....

(3 marks)

(b) Seafood and rice are both high-risk foods.

(i) Explain why seafood is a high-risk food.

....

....

....

(3 marks)

(ii) Give an example of a health and safety procedure that should be followed when serving this dish.

....

....

(1 mark)

(iii) Complete the table below by describing two safety and hygiene procedures the restaurant should follow when storing and cooking seafood.

Storing seafood	....
Cooking seafood	....

(4 marks)

(c) The restaurant staff want to buy some electrical equipment to help them prepare the seafood dish. For each of the processes given below, suggest one piece of electrical equipment that they could use. Give two reasons for each of your suggestions.

(i) Chopping the vegetables.

....

....

....

....

(3 marks)

(ii) Making a tomato purée used in the sweet and sour sauce.

....

....

....

....

(3 marks)

17

6 Look at the picture below, which shows the inner and outer packaging of a breakfast cereal.

(a) (i) Name the material that the outer box is made from.

....

(1 mark)

inner packet outer box

(ii) Describe the advantages and disadvantages of using this material to package the product.

Advantages: ....

....

....

Disadvantages: ....

....

(4 marks)

(iii) Name the material that the inner packet is made from.

....

(1 mark)

Leave blank

(b) The sustainability of product packaging must be considered during product design. Evaluate the sustainability of the packaging materials you named in (a).

....

....

....

....

....

....

(4 marks)

(c) Foods that are GM or that contain more than 1% GM ingredients must be clearly labelled on the packaging by law.

(i) Briefly describe what is meant by a GM food.

....

....

....

(2 marks)

(ii) Evaluate the advantages and disadvantages of producing GM foods.

....

....

....

....

....

....

....

....

....

....

(6 marks)

18

END OF QUESTIONS

Section One — The Design Process

Page 15 (Warm-up Questions)

1 a) The group of people you want to sell your product to.

b) Any reasonable answers, e.g. gender, age, job, hobbies, lifestyle, income.

c) Any five sensible answers, e.g. you could ask about age, job, favourite fruits and/or vegetables, whether they like salad dressings, where/when they eat salads.

2 Any sensible answers, e.g.

a) Do you like cheese in your sandwiches?

b) What kind of fillings do you like?

c) Which of the following types of bread do you like?
White; brown; wholemeal; rye.

3 Any three reasonable answers, e.g. shape, colour, texture, new ingredient, recipe, production process, product packaging. (Other answers are possible.)

4 Any two sensible suggestions, e.g. try a different kind of dough for the pizza base, use fresher toppings, add olive oil.

5 Answer should look something like the following:

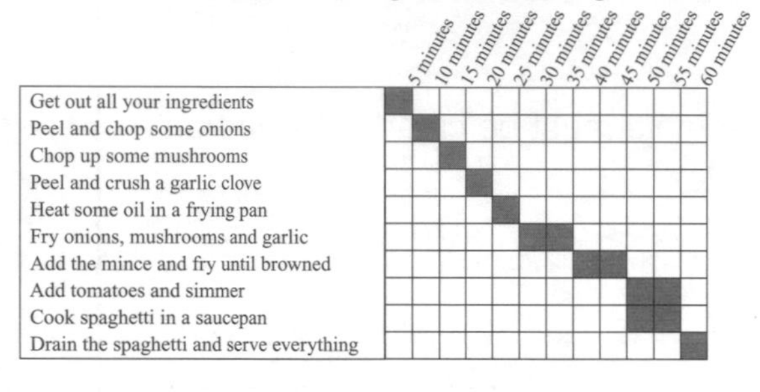

Page 17-18 (Exam Questions)

1 (a) Notes / annotated sketches should both show:

- a savoury meal for one person ***[1 mark]***.
- a tomato-based sauce, e.g. tomato and herb ***[1 mark]***.
- a high carbohydrate meal that uses pasta ***[1 mark]***.
- a meal high in protein, e.g. uses meat or an alternative source of protein such as beans or lentils ***[1 mark]***.
- a product that has sensory appeal, e.g. a nice flavour, texture, aroma ***[1 mark]***.
[Maximum of 5 marks available for each design.]

E.g. spaghetti with meatballs:

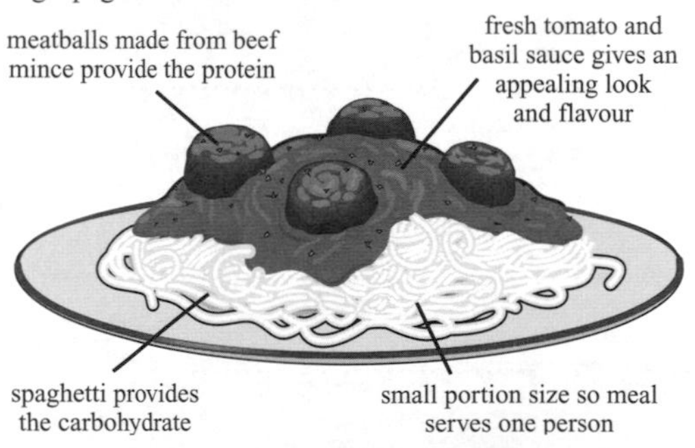

E.g. vegetarian lasagne:

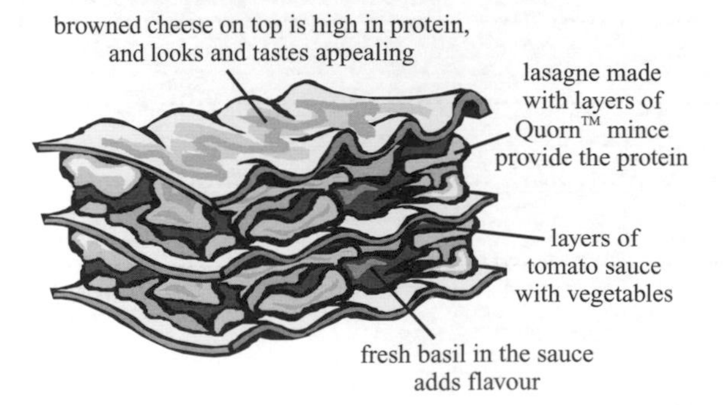

(b) Plan should show:

- A clear, logical order ***[1 mark]***.
- Safety control checks, e.g. personal/kitchen/food hygiene ***[1 mark]***.
- Feedback from safety control checks ***[1 mark]***.
- Key times ***[1 mark]***.
- Key temperatures ***[1 mark]***.

E.g. flowchart:

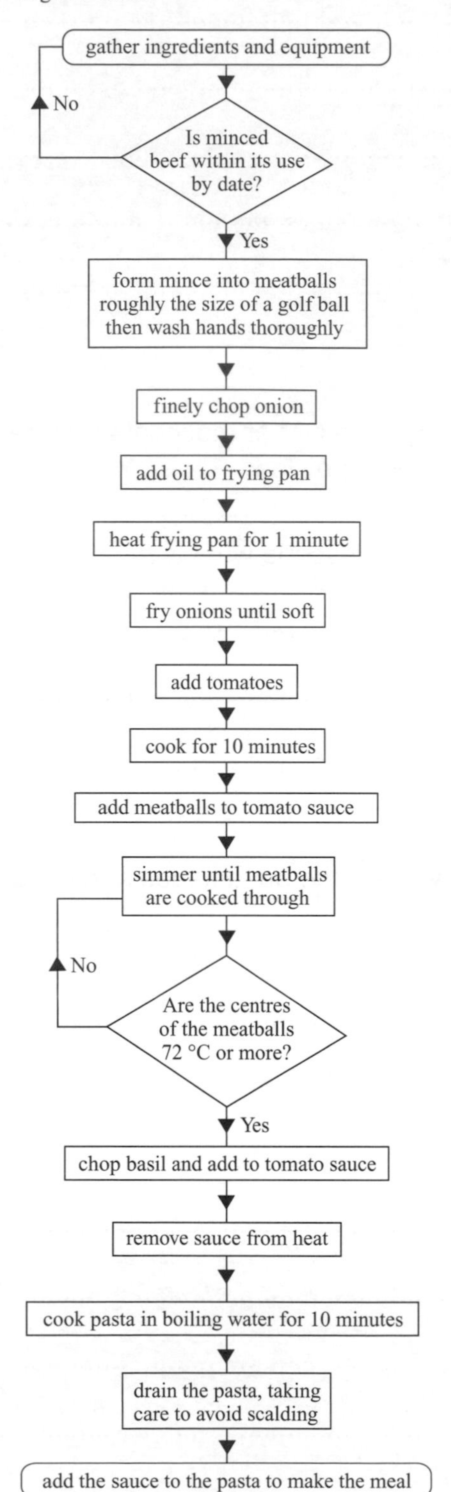

Section Two — Properties of Food

Page 33 (Warm-up Questions)

1 a) It acts as a preservative.
   b) It speeds up fermentation.
   c) It adds sweetness and colour.
2 To tenderise the meat (by partly breaking down the fibres).
3 E.g. 1. Always cook eggs thoroughly. 2. Make sure chicken is cooked properly. 3. Use dried or pasteurised eggs.
4 E.g. A diet with no fat would be unhealthy as the person would get less vitamins and fatty acids, which are needed to keep the body healthy.
5 Sam should eat more foods that contain calcium, e.g. milk - this will make his bones stronger. He should also eat more fruit and veg, as these contain vitamin C that will help his body fight infections. (Other answers are possible but they should focus on naming foods that will help with the particular problems mentioned in the question.)

Page 34 (Exam Questions)

1 (a) Strong flour ***[1 mark]***.
   (b) Gluten ***[1 mark]***.
   (c) To make the dough rise / to produce carbon dioxide ***[1 mark]***.
2 Notes / sketch and annotations should show a suitably adapted sponge based product.
   3 marks for 3 different ways of cutting calories, e.g:
   - less cream used / low-fat cream used ***[1 mark]***.
   - low-sugar jam used / jam replaced with fresh fruit ***[1 mark]***.
   - portion size reduced ***[1 mark]***.
   - reduced fat/sugar sponge cake ***[1 mark]***.
   - sugar substitute used that is lower in calories ***[1 mark]***.
3 (a) Protein helps the body to grow ***[1 mark]*** and to repair muscles and tissues ***[1 mark]***. Proteins are made of amino acids, and some proteins contain essential amino acids which the body can't make itself ***[1 mark]***.
   (b) E.g. fat ***[1 mark]*** / vitamin A/B2/D ***[1 mark]*** / iodine ***[1 mark]***.

Page 43 (Warm-up Questions)

1 a) MSG (monosodium glutamate).
   b) A sweetening agent, e.g. sugar or saccharin.
2 A sharp, sour flavour.
3 a) She could replace the beef with another source of protein, e.g., lentils or Quorn™. Or she could make a vegetable lasagne instead.
   b) Her friend can't tolerate eating gluten, which is found in wheat so she can't eat normal pasta. Susie could use gluten-free pasta sheets in her lasagne.
4 A food that has been artificially changed to have useful characteristics by having its genes altered.

Page 44 (Exam Questions)

1 (a) Any two of: include fruit and vegetables in the diet ***[1 mark]***. / Reduce the amount of fatty foods eaten ***[1 mark]***. / Reduce the amount of sugary foods eaten ***[1 mark]***. / Eat more starchy foods ***[1 mark]***.
   (b) Chocolate bar ***[1 mark]***.
2 The additive has passed a safety test ***[1 mark]*** and can be used throughout the European Union ***[1 mark]***.
3 (a) A food that has been artificially modified ***[1 mark]*** to provide a particular health benefit, on top of its normal nutritional value ***[1 mark]***.
   (b) E.g. any two of: fruit juices with added calcium ***[1 mark]***. / Eggs with added omega-3 ***[1 mark]***. / Golden Rice/bread with extra vitamins ***[1 mark]***. / Margarine with added omega-3/vitamins ***[1 mark]***.
   (c) Functional foods mean people with a poor diet can easily eat more of a certain nutrient ***[1 mark]***. People who can't eat particular foods can get the nutrients they lack by eating functional foods ***[1 mark]***. Some functional foods could help solve some health problems caused by malnutrition in poor countries ***[1 mark]***.

Section Three — Food Processes

Page 59 (Warm-up Questions)

1 a) egg yolk (lecithin)
   b) jelly / mousse / cheesecake / jam
2 a) To improve the taste of food / to bring out the flavour of food.
   b) Any suitable example, e.g. marinate meat in barbecue marinade before cooking to add flavour to the meat.
3 You could use folding when making a cake — you use a spoon or a spatula to fold the mixture in half repeatedly. This stops air being lost when mixing, so helps the cake to rise.
4 Stir-frying because food is cooked quickest using the least amount of oil, so less fat is absorbed by the food.
5 Boiling is quite a harsh method of cooking and the bubbles would probably break up the fish, so it'd fall apart. You could boil the fish in a bag.
6 E.g. you could decorate the cake with icing, add marzipan figures, etc.

Page 60 (Exam Questions)

1 (a) E.g. using fresh products may give a better flavour ***[1 mark]***.
   (b) E.g. any three of: the manufacturer can't pick and choose exactly what they want ***[1 mark]***. / Late delivery from the supplier will hold up the production line ***[1 mark]***. / Extra space might be needed to store a bulk buy of standard components ***[1 mark]***. / There's extra packing and transport involved so it might be bad for the environment/be more expensive ***[1 mark]***.
2 (a) (i) Conduction ***[1 mark]***.
      (ii) Radiation ***[1 mark]***.
   (b) *Advantage* — e.g. food cooks quickly at a high temperature ***[1 mark]***. / It's fairly healthy as fat drips off the food ***[1 mark]***. / The outside of the food looks and tastes nice and can have a crispy texture ***[1 mark]***.
   Disadvantage — e.g. it's hard to tell if the food is cooked all the way through because the outside cooks much quicker than the inside ***[1 mark]***. / using high temperatures makes it easier to burn the food ***[1 mark]***.
3 E.g. any three of: only boil the amount of water you need ***[1 mark]*** because boiling water in a kettle or pan uses a lot of energy ***[1 mark]***. / use an appropriate size of pan and ring ***[1 mark]*** because a lot of energy is lost and cooking will take longer if you use a pan or ring that's too big ***[1 mark]*** / cover saucepans and pots with lids ***[1 mark]*** because this stops heat being lost to the kitchen ***[1 mark]***. / cook different foods together ***[1 mark]*** because this reduces the number of rings that are used ***[1 mark]***. / use the most energy efficient cooking method ***[1 mark]***, for example, a microwave ***[1 mark]***.

Page 72 (Worked Exam Questions)

1 a) Batch production — because lots of sandwiches can be made in one go so it's quick, and you can change between making batches of different kinds of sandwiches.
   b) One-off production / jobbing production.
2 Production stops while the problem is investigated and solved. E.g. the weighing scales may be slightly out so the flapjacks are ending up slightly smaller. The weighing scales would be re-set and production restarted.
3 a) Chicken/sauce/eggs. They're high-risk foods because they're moist and high in protein, so bacteria grow quickly in them.
   b) Any three from: keep the knives and the chopping boards separate from anything else he's preparing / wash his hands thoroughly after handling raw meat/eggs/sauce / never store raw meat and cooked meat together / don't let the blood and juices of raw meat drip onto other food.

Page 73 (Exam Questions)

1 (a) Any three of: sickness *[1 mark]* / diarrhoea *[1 mark]* / stomach cramps *[1 mark]* / fever *[1 mark]*.

(b) Heating food to over 72 °C should kill most of the bacteria *[1 mark]*.

2 (a) (i) 0 and 5 °C *[1 mark]*.

(ii) Because chilling slows the growth of bacteria, so food is safer for longer *[1 mark]*.

(b) *Advantages* — e.g. freezing food greatly extends its shelf life *[1 mark]*. / Frozen food keeps all its nutrients (unlike heating methods of preservation) *[1 mark]*. / Freezing food stops bacteria growing/makes bacteria dormant *[1 mark]*. / Freezing food helps it to keep its colour and flavour *[1 mark]*.
***[1 mark for each advantage, up to 2 marks]*.**
Disadvantages — e.g. freezing food doesn't kill the bacteria *[1 mark]*. Freezing food can change its texture *[1 mark]*.
***[1 mark for each disadvantage, up to 2 marks]*.**

3 E.g. computerised weighing equipment is quicker and easier to use than weighing things non-electrically, e.g. using a balance pan *[1 mark]*. Products are consistent *[1 mark]* because they can be accurately weighed with less room for human error, e.g. to within 0.05 g *[1 mark]*. Scales can be preset to weigh different ingredients, which saves time *[1 mark]*.

Section Four — Marketing and the Environment

Page 87 (Warm-up Questions)

1 a) Any type of luxury product that's ready to eat or quick to cook, e.g. smoked pheasant salad.

b) The high quality/luxurious nature of the product may appeal to rich people. The speed of preparation should appeal to busy people.

2 It's a movement to help workers in developing countries get a fair price for their produce and good working conditions.
Advantages: any one of, e.g. farmers and workers are treated fairly, help workers to invest in their communities, workers have a better quality of life.
Disadvantages: e.g. excess food causes world prices to fall, so those not in a fair trade scheme might lose out.

3 a) E.g. use stale bread to make bread-and-butter pudding

b) E.g. use sugar beet waste to feed pigs.

4 a) modified atmosphere packaging (MAP)

b) The food is put into plastic packaging with a mixture of oxygen, nitrogen and carbon dioxide in specific proportions. It's then sealed and chilled.

Page 88 (Exam Questions)

1 *Fruit salad* — e.g. people trying to lose weight *[1 mark]* as they need to eat low-fat foods *[1 mark]*. / Elderly people *[1 mark]* as they may need to cut down on fats in their diet *[1 mark]*. / People concerned with healthy eating *[1 mark]* as it will contain some of their five a day *[1 mark]*.

High energy snack bar — e.g. athletes *[1 mark]* as they need food that provides lots of energy *[1 mark]*. / People with active jobs *[1 mark]* as they need food that provides lots of energy *[1 mark]*.

Calcium enriched cereal — e.g. pregnant women *[1 mark]* as they need extra calcium for healthy baby development *[1 mark]*. / Toddlers/children *[1 mark]* as they need calcium for growth and development *[1 mark]*.

2 (a) 33.0 g *[1 mark]*.

(b) This isn't a reasonable claim *[1 mark]* because the nutritional information shows that the snack contains 33.0 g of fat per 100 g / the snack is nearly a third fat *[1 mark]*.

3 E.g. any three of: to keep the product together *[1 mark]*. / To protect the product from being damaged whilst it's being transported/displayed/stored *[1 mark]*. / To preserve the product *[1 mark]*. / To avoid contamination *[1 mark]*. / To identify what the product is *[1 mark]*. / To give customers useful information *[1 mark]*.

Practice Exam Paper

1 (a) Notes / annotated sketches should both show:
- a pastry-based product *[1 mark]*.
- a savoury filling *[1 mark]*.
- suitability for vegetarians, e.g. no meat used *[1 mark]*.
- have a suitable glazed finish, e.g. brushed with egg *[1 mark]*.
- a product that provides sensory appeal, e.g. flavour, texture, aroma *[1 mark]*.
- clear sketching and good communication of details *[1 mark]*

[Maximum of 6 marks available for each design.]

E.g. mushroom and leek pie:

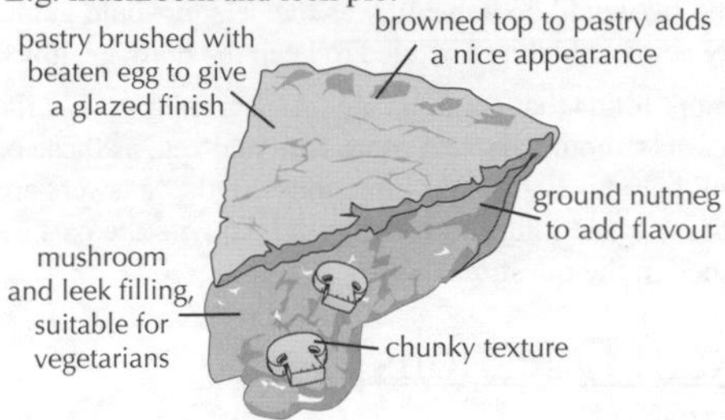

E.g. mushroom and asparagus quiche:

(b) Flowchart should show:
- A clear, logical order *[1 mark]*.
- Correct terminology, e.g. 'rub butter' *[1 mark]*.
- Quality control checks, e.g. size, shape *[1 mark]*.
- Safety control checks, e.g. personal/kitchen/food hygiene *[1 mark]*.
- Feedback from control checks *[1 mark]*.
- Key times *[1 mark]*.
- Key temperatures *[1 mark]*.
- Finishing techniques *[1 mark]*.

E.g. flowchart:

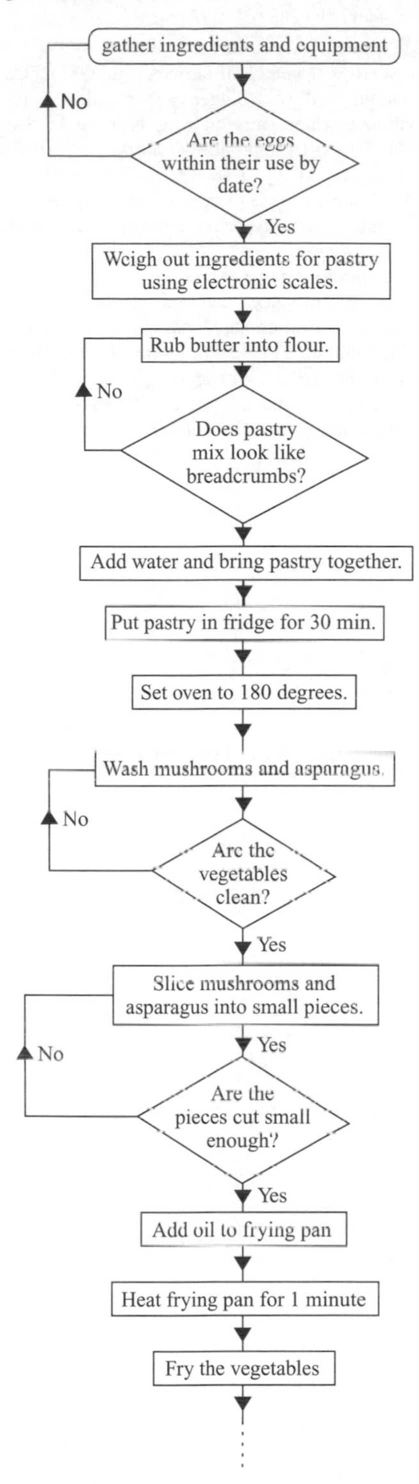

(c) Any vitamins, relevant ingredient and function.

E.g. *Vitamin*: Vitamin C (ascorbic acid) ***[1 mark]***: *Ingredient* — asparagus ***[1 mark]***. *Function* — protects the body from infection/allergies / helps in the absorption of calcium and iron from food / helps to keep blood vessels healthy / helps to heal wounds ***[1 mark]***.

Vitamin: Vitamin A ***[1 mark]***: *Ingredient* — eggs ***[1 mark]***. *Function* — needed for good eyesight (especially night vision) / needed for growth and function of tissues ***[1 mark]***.

(d) E.g. the results show consumers thought that puff pastry has a better texture, but flaky pastry has a better flavour, moistness and overall appearance ***[1 mark]***. Both pastries scored the same for their colour ***[1 mark]***. The flaky pastry scored higher in most of the categories so is probably more suitable to use for my pastry product because it will be more popular with consumers ***[1 mark]***. You could use the results to further improve the flaky pastry, e.g. improve its texture, or compare the results with previous/future sensory tests with target consumers ***[1 mark]***

2 (a) Coagulation — eggs become more solid at high temperatures ***[1 mark]***, then set and stay thickened ***[1 mark]***.

Emulsification — egg yolks contain lecithin ***[1 mark]***, which keeps emulsions stable / stops emulsions from separating ***[1 mark]***.

Aeration — e.g. when egg white is beaten, the protein in it is stretched ***[1 mark]*** and air becomes trapped ***[1 mark]***.

(b) (i) Caster sugar ***[1 mark]*** — e.g. because it will give the sponge a smooth texture / it has finer granules ***[1 mark]***.

(ii) Icing sugar ***[1 mark]*** — e.g. because it instantly dissolves in water to form a smooth paste / it's a fine powder so it can be dusted over cakes easily ***[1 mark]***.

(c) (i) Bicarbonate of soda ***[1 mark]***.

(ii) It breaks down when heated to produce carbon dioxide ***[1 mark]***. The carbon dioxide bubbles expand and make mixtures rise ***[1 mark]***.

(iii) Alkalis have an unpleasant, bitter taste ***[1 mark]***.

(iv) Any two strongly flavoured products, e.g. gingerbread ***[1 mark]***, chocolate cake ***[1 mark]***. The taste of the alkali would not be a problem because the strong flavour of the product would mask the unpleasant taste ***[1 mark]***.

3 (a) *Pasta* — starch acts as a bulking agent ***[1 mark]***. The starch granules swell when a liquid is added ***[1 mark]***.

White sauce — starch acts as a thickening agent ***[1 mark]***. When heat is applied to a starch and liquid mixture, gelatinisation occurs and thickens the sauce ***[1 mark]***.

(b) (i) The tuna could be replaced with tofu / Quorn™ / more vegetables ***[1 mark]***. It could be served with no cheese on the top ***[1 mark]***. It could be made without white sauce / the white sauce could be made with a milk alternative/soya milk ***[1 mark]***.

(ii) The pasta could be replaced with gluten-free pasta ***[1 mark]***.

(iii) The product could be prepared using ingredients that do not contain nuts or traces of nuts ***[1 mark]***.

(c) (i) Lactose intolerance ***[1 mark]***.

(ii) The white sauce ***[1 mark]*** and grated cheese ***[1 mark]***.

(d) (i) The product is suitable for vegetarians/meat free ***[1 mark]***. E.g. it helps people who do not eat animal products because of personal/religious beliefs to know if the food is suitable for them ***[1 mark]***. It helps people who are allergic to animal products to know if the food is safe for them to eat ***[1 mark]***.

(ii) The product may contain nuts ***[1 mark]***. E.g. any two of: it lets people who are allergic to nuts know that the product may not be safe for them to eat ***[1 mark]***. / It lets consumers know that the product has been made in an area where there are nuts, so it could be contaminated ***[1 mark]***. / It lets consumers know that nuts or nut traces may be in the product even if they're not listed in the ingredients ***[1 mark]***.

(iii) Some or all of the packaging can be recycled ***[1 mark]***. E.g. it reminds consumers to recycle the packaging ***[1 mark]***. It helps consumers to dispose of the packaging in a suitable way ***[1 mark]***.

4 (a) (i) E.g. it saves the manufacturer time because they don't have to prepare the tomato sauce first ***[1 mark]***. / It saves the manufacturer money because it is more cost-effective to buy the sauce in bulk ***[1 mark]***. / It makes the product more likely to be consistent because the sauce used will all be the same colour, texture, etc. ***[1 mark]***. / Less machinery/equipment/staff are needed by the manufacturer because the sauce is already prepared ***[1 mark]***. ***[1 mark for each advantage, up to 3 marks.]***

(ii) E.g. adding more salt ***[1 mark]***. Adding some herbs or spices ***[1 mark]***.

(iii) Pre-made pizza bases ***[1 mark]***.

(b) E.g. using standard components involves using extra packaging ***[1 mark]***, which uses up a lot of resources, e.g. oil for plastic ***[1 mark]***. It involves extra transport ***[1 mark]***, which means more fossil fuels are burnt / more carbon dioxide is released / global warming is contributed to ***[1 mark]***.

(c) (i) Pizza sales decreased between 2005 and 2007 ***[1 mark]***, but then dramatically increased between 2007 and 2009 ***[1 mark]***.

(ii) E.g. People may save money on eating out in restaurants but instead treat themselves with a take-away pizza, which is cheaper ***[1 mark]***.

(d) *CAD examples*, e.g. designing the appearance/packaging of the pizza ***[1 mark]***. / Modelling portion sizes ***[1 mark]***. / Calculating the nutritional content/cost/profit/shelf life ***[1 mark]***. / Analysing sensory data ***[1 mark]***. / Presenting information ***[1 mark]***. / Showing the assembly procedures ***[1 mark]***.
[1 mark for each CAD example, up to two marks.]
CAM examples, e.g. monitoring/changing the production process ***[1 mark]***. / Using computer controlled equipment, e.g. electronic scales ***[1 mark]***. / Weighing out ingredients ***[1 mark]***. / Setting oven temperature/cooking time ***[1 mark]***.
[1 mark for each CAM example, up to 2 marks.]

5 (a) E.g. any three of: check ingredients are bought from a reliable supplier ***[1 mark]***. / Visually check the condition of the ingredients/packaging ***[1 mark]***. / Check ingredients have been stored at the correct temperature ***[1 mark]***. / Check the use by/sell by/best before dates to make sure the ingredients are fresh ***[1 mark]***. / Check the weights using computers/ digital scales ***[1 mark]***. / Check for any physical/biological/chemical contamination ***[1 mark]***.

(b) (i) Seafood is moist ***[1 mark]*** and high in protein ***[1 mark]***, so bacteria can grow/multiply very quickly ***[1 mark]***.

(ii) E.g. after the seafood has been cooked it should be served straight away ***[1 mark]***.

(iii) *Storing seafood* — follow the storage instructions, e.g. store between 0 to 5 °C ***[1 mark]***. / Rotate stock/use old purchases first, before they go out of date ***[1 mark]***. / Keep seafood sealed or covered up ***[1 mark]***. / Keep raw seafood away from other foods ***[1 mark]***.
Cooking seafood — ensure the centre of the seafood is heated to 72 °C ***[1 mark]***. / Cook for the correct amount of time ***[1 mark]***. / Make sure the seafood is cooked all the way through ***[1 mark]***.
[Maximum of 2 marks for each of storing seafood and cooking seafood].

(c) (i) Food processor ***[1 mark]***. This will chop the vegetables consistently/the same every time ***[1 mark]***. / It saves time/effort of chopping by hand ***[1 mark]***. / It's more hygienic than doing it by hand ***[1 mark]***. / It's safer than using a sharp knife ***[1 mark]***.
[1 mark for naming electrical equipment and up to 2 marks for relevant reasons].

(b) Blender ***[1 mark]***. This will mix the ingredients together to get a smooth result ***[1 mark]***. / It saves time/effort of doing it by hand ***[1 mark]***. / It gets a consistent product using the same settings ***[1 mark]***.
[1 mark for naming electrical equipment and up to 2 marks for relevant reasons].

6 (a) (i) Cardboard ***[1 mark]***.

(ii) *Advantages* — e.g. lightweight ***[1 mark]***. / Flexible ***[1 mark]***. / Easy to print on ***[1 mark]***. / Biodegradable ***[1 mark]***. / Easy/cheap to recycle ***[1 mark]***.
[1 mark for each advantage, up to 2 marks].

Disadvantages — e.g. you can't see the contents ***[1 mark]***. / It's not very rigid so could get squashed ***[1 mark]***.
[1 mark for each disadvantage, up to 2 marks].

(iii) Plastic ***[1 mark]***.

(b) E.g. Both cardboard and plastic can be recycled ***[1 mark]***. Cardboard is biodegradable but plastic is not ***[1 mark]***. Cardboard is made using a renewable resource (trees) whilst plastic comes from a non-renewable resource (oil) ***[1 mark]***. So cardboard is a more sustainable material than plastic ***[1 mark]***.

(c) (i) A genetically modified food is one that's had its genes altered ***[1 mark]*** to give it useful characteristics ***[1 mark]***.

(ii) *Advantages* — e.g. farmers can use GM crops that will grow quicker than normal crops ***[1 mark]***. If farmers plant GM maize that is pest-resistant, they will get a bigger yield of maize because less of the crop will be eaten or damaged by pests ***[1 mark]***. This makes it cheaper for the farmer to produce ***[1 mark]*** and so makes it cheaper for the consumer to buy ***[1 mark]***. The consumer benefits because foods can be made to ripen earlier in the year, when they wouldn't normally be available ***[1 mark]***, and their shelf life can be increased ***[1 mark]***.
[1 mark for each advantage, up to 3 marks].
Disadvantages — e.g. GM producers can't sell their food everywhere because of EU restrictions ***[1 mark]***. The long-term health effects of GM foods aren't yet known ***[1 mark]***, and there are concerns that modified genes could get out into the wider environment and cause environmental problems ***[1 mark]***.
[1 mark for each disadvantage, up to 3 marks].

Index

Index

Index

Q

R

S

T

U

V

W

Y